Lectures on
Financial Mathematics:
Discrete Asset Pricing

Synthesis Lectures on Mathematics and Statistics

Editor
Steven G. Krantz, *Washington University, St. Louis*

Lectures on Financial Mathematics: Discrete Asset Pricing
Greg Anderson and Alec N. Kercheval
2010

Jordan Canonical Form: Theory and Practice
Steven H. Weintraub
2009

The Geometry of Walker Manifolds
Miguel Brozos-Vázquez, Eduardo García-Río, Peter Gilkey, Stana Nikcevic, and Rámon Vázquez-Lorenzo
2009

An Introduction to Multivariable Mathematics
Leon Simon
2008

Jordan Canonical Form: Application to Differential Equations
Steven H. Weintraub
2008

Statistics is Easy!
Dennis Shasha and Manda Wilson
2008

A Gyrovector Space Approach to Hyperbolic Geometry
Abraham Albert Ungar
2008

Lectures on Financial Mathematics: Discrete Asset Pricing

Greg Anderson and Alec N. Kercheval

ISBN: 978-3-031-01271-6   paperback
ISBN: 978-3-031-02399-6   ebook

DOI 10.1007/978-3-031-02399-6

A Publication in the Springer series
SYNTHESIS LECTURES ON MATHEMATICS AND STATISTICS

Lecture #7
Series Editor: Steven G. Krantz, *Washington University, St. Louis*
Series ISSN
Synthesis Lectures on Mathematics and Statistics
Print 1938-1743   Electronic 1938-1751

Lectures on
Financial Mathematics:
Discrete Asset Pricing

Greg Anderson
Bank of America Merrill Lynch

Alec N. Kercheval
Florida State University

SYNTHESIS LECTURES ON MATHEMATICS AND STATISTICS #7

ABSTRACT

This is a short book on the fundamental concepts of the no-arbitrage theory of pricing financial derivatives. Its scope is limited to the general discrete setting of models for which the set of possible states is finite and so is the set of possible trading times – this includes the popular binomial tree model. This setting has the advantage of being fairly general while not requiring a sophisticated understanding of analysis at the graduate level.

Topics include understanding the several variants of "arbitrage", the fundamental theorems of asset pricing in terms of martingale measures, and applications to forwards and futures.

The authors' motivation is to present the material in a way that clarifies as much as possible why the often confusing basic facts are true. Therefore the ideas are organized from a mathematical point of view with the emphasis on understanding exactly what is under the hood and how it works. Every effort is made to include complete explanations and proofs, and the reader is encouraged to work through the exercises throughout the book.

The intended audience is students and other readers who have an undergraduate background in mathematics, including exposure to linear algebra, some advanced calculus, and basic probability. The book has been used in earlier forms with students in the MS program in Financial Mathematics at Florida State University, and is a suitable text for students at that level. Students who seek a second look at these topics may also find this book useful.

KEYWORDS

arbitrage, martingale, incomplete markets, forward measure, forward contract, futures, tree models

Contents

Preface

This is a short book on discrete asset pricing theory, the way we think it should be done. It is based on course notes used for several years at Florida State University, in evolving form, for Masters degree students in Financial Mathematics.

The intended audience is students and other readers who have mathematical training at the undergraduate level, including linear algebra, advanced calculus, and some probability. Some exposure to measure theory would be useful, but an introduction to the needed concepts is supplied in Appendix A of this book: Probability Refresher.

We do not assume readers have had graduate-level training in probability or real analysis, which makes this book more elementary than many of the growing number of mathematical finance texts becoming available. However, we do adopt a mathematician's point of view, so readers will find the book more mathematical in flavor than many elementary finance or economics books on the same topic.

Our motivation is to try to present the material in a way that clarifies as much as possible why the often confusing basic facts are true. Since understanding "why" is the primary vocation of the mathematician, we have organized the ideas from a mathematical point of view in an effort to expose as clearly as possible what is under the hood and how it works.

By "discrete asset pricing theory" we mean the no-arbitrage theory of derivative pricing for models with a finite set of possible states and a finite set of possible trading times, such as the well-known binomial tree model of Cox-Ross-Rubinstein [CRR79]. We feel that the customary study of the binomial case obscures some of what is really going on, so we study the more general discrete case in this book.

The discrete theory is a companion to the continuous no-arbitrage theory as exemplified by the Black-Scholes model pricing options in continuous time. This is also an important case, but it requires more advanced mathematical tools than does the finite case; often the depth of the mathematics can make it hard to recognize the essence of the financial questions, including what assumptions are being made at any given time. This helps motivate our focus on the discrete case where we can hope to come to a complete understanding of what is going on without requiring mathematical sophistication at the PhD level.

Still, readers will get the most from this book if they are prepared to dig into the reasons – that is, to read and write proofs. Many of the claims are relegated to numbered Exercises interspersed with the text. The reader is strongly encouraged to do these exercises when they appear.

The core of this book centers on the fundamental theorems of asset pricing, treated in Chapter 3. These theorems describe the relationship between the absence of arbitrage and the existence of a "martingale measure" whose role is to define a simple pricing formula for derivative contracts, like options. These ideas were first developed by Harrison and Pliska [HP81] as part of a series of works by them and Kreps (including [HK79] and [Kre81]) developing a general framework for no-arbitrage arguments. They built on the earlier work of Black, Scholes, and Merton ([BS73], [Mer73]), who pioneered the idea of using no-arbitrage as a pricing principle in mathematical models of asset prices.

Since 1981, understanding of the fundamental theorems of asset pricing in more general contexts, such as in continuous time models like the Black-Scholes framework, progressed steadily and reached a pinnacle with the 1990's work of Delbaen and Schachermayer ([DS94], [DS98]). An excellent reference for the full story is their book [DS06].

Finite models, the topic of this book, are vastly simpler, which has certain advantages for the beginner. Computations can be done in a fairly elementary way that exposes the underlying concepts in their simplest forms, and they may clarify for the non-expert certain ideas like the forward measure, why futures and forwards are different, the nature of the convexity correction, the difference between complete and incomplete markets, etc. Moreover, since finite models are commonly used as computational tools, a clear understanding of them is essential for practitioners.

Fifteen years ago, there were not many books on this topic, but the intervening years have seen a boom in financial mathematics books at many levels. Popular books that treat our topic from varying viewpoints include Hull [Hul08], Baxter and Rennie [BR96], Duffie [Duf96], Pliska [Pli97], Elliot and Kopp [EK99], Shreve [Shr04], Delbaen and Schachermayer [DS06], and this list is by no means complete.

At Florida State University, the second author has used parts of this book in the form of lecture notes to cover about one-third of a Master's-level semester course on Financial Engineering as a prelude to the study of continuous-time models like the Black-Scholes model. A rigorous treatment of the discrete case for students who have not yet mastered graduate level real analysis is a great advantage in motivating a subsequent and more informal treatment of continuous-time models for these students.

We have been guided in writing this book to focus primarily on topics that illuminate the central ideas of no-arbitrage pricing, rather than on producing a complete reference book or a course text covering a full semester's worth of topics. In doing so, we hope to provide readers with a slim but inviting volume that can be worked through as a coherent story and used as a jumping-off point to further study. We also hope it will be used by readers who have been exposed to the basic ideas before but who seek a second treatment that might illuminate some of the puzzling questions that sometimes remain after a first pass.

Chapter 1 introduces the key ideas in the simplest context of a single time step. Careful attention to the details here will illuminate what is going on when we add the passage of time to the model in Chapter 2. This second chapter treats the general discrete case, including the central definitions and a careful look at the various flavors of arbitrage, which drive all our pricing formulas.

Chapter 3 is the center of this book and describes three fundamental theorems of asset pricing, which are about martingale measures, pricing formulas, and their relation to the no arbitrage condition. The summit of the chapter is the proof of the second fundamental theorem.

Chapter 4 on Forwards and Futures is a natural sequel to the second fundamental theorem of Chapter 3. The pivotal measure in that proof is identified as the forward measure, which can be understood in terms of forward prices; the martingale measure has a corresponding role with futures prices. This chapter also includes some computational examples illustrating the pricing of options, forwards, and futures in discrete models with any finite number of assets.

Chapter 5 looks carefully at the case of incomplete markets, and how the no-arbitrage condition restricts the possible prices of a claim.

Beginners may wish to start with the first Appendix: Probability Refresher. This summarizes most of the needed probability ideas for this book, with a focus on sigma-algebras and their interpretation. These are the central ideas in our intended application and most likely to be still unfamiliar to the beginner. We have largely confined ourselves to the simplest case of finite sample spaces – which has both the virtue and drawback of bypassing all the subtlety of convergence questions and functions spaces necessary for the more general case. It is hoped this appendix is mostly review for the typical reader, though we have tried to be complete enough for the first-timer, who will nevertheless profit from consulting other texts on probability, such as Jacod and Protter [JP00], Chung [Chu74], or Billingsley [Bil95].

The final appendix contains the proof of a geometric fact used in the second fundamental theorem of asset pricing of Chapter 3. This is elementary but may be omitted on first reading.

Greg Anderson and Alec N. Kercheval
August 2010

CHAPTER 1

Overture: Single-Period Models

1.1 WHAT IS A MODEL?

We don't know the future: we know neither all the possible future outcomes, nor the true likelihood of any single one. Nevertheless, we need to plan ahead, so we make provisional models of the uncertain future, in hope they will be useful. Financial asset pricing is based on such models.

A mathematical asset pricing model takes the following form: we specify some precise set of futures states of the world, along with a probability distribution (or class of distributions) for those states. In each state, we assign potential asset prices to our securities. These prices could come via an endogenous market equilibrium derived from the utility functions of traders, or they can be imposed exogenously as inputs to the problem.

In such models, our fuzzy and limited knowledge of the future is replaced by a well-specified set of states, and our vast uncertainty is replaced by the very limited uncertainty of not knowing which of the specified outcomes will occur in advance. Still, there's something to learn by living for a while in the artificial world of a pricing model, and that is the subject of this book.

1.2 WARM-UP: A FORWARD CONTRACT

Imagine we are in a market with a stock paying no dividends but with an uncertain future price S_t at each time t. We also have a bank account paying fixed interest rate r (continuously compounded), and we are free to borrow or lend to the bank account and take long or short positions in the stock.

At $t = 0$ (today), we want to enter into a contract to purchase one share of the stock, now worth S_0, at future maturity time $t = T$. We need to decide today what strike price K, to write into the contract, which is the price we will pay at $t = T$ in exchange for the stock, then worth S_T. The net payoff is $S_T - K$, which is unknown at time $t = 0$.

What is the fair strike price K to set in this contract? Fair means that financially we would be indifferent to taking either side of the contract or declining to participate; equivalently, it means the value of the resulting contract is zero. The strike price K such that the contract has value zero, is called the "forward price" of the stock.

It is tempting to think the forward price should be the one that makes the net value of the payoff zero on average, or $K = E[S_T]$, where E is the expectation with respect to the statistical probability measure describing the possible values of the random variable S_T. For example, if there is a 90% chance that $S_T = \$10$, and a 10% chance that $S_T = \$5$, then we might expect the forward price should be $9.50.

This is intuitively appealing but wrong. Consider the following strategy:

- at time $t = 0$: enter into a short position in the forward contract with strike K (that is, contract to deliver the stock), borrow S_0 dollars from the bank, and purchase 1 share of the stock.

- at time $t = T$: deliver the stock to your contract counter-party, receive K dollars in exchange, and pay off the loan amount, $S_0 e^{rT}$.

Your net profit is $K - S_0 e^{rT}$. If this is positive, it represents a riskless profit because the above strategy has zero buy-in cost. Such a thing cannot exist in a stable market. Therefore, we must have $K \leq S_0 e^{rT}$. By reversing the above positions, we can also argue we must have the reverse inequality, so $K = S_0 e^{rT}$ is the only possible strike price that avoids a destabilizing arbitrage (riskless profit by one side).

It may at first seem counter-intuitive that the correct forward price $S_0 e^{rT}$ is unrelated to our original guess $E[S_T]$ and so is unrelated to the probability distribution of the stock price S_T. This is the theme of arbitrage pricing theory: prices must conform to the requirement of no arbitrage first, and only when that requirement is satisfied can expectations or preferences enter.

With this motivation, we now carefully develop our pricing theory from the beginning. We will return to forward contracts again in Chapter 4.

1.3   A SINGLE TIME PERIOD

All our asset pricing models address the following question: what is the value today of a specified uncertain future payment? To understand what makes these models tick, we need to see how the simplest ideas are reflected in even the most complex models, so we start with the simplest possible model and work our way up.

Let's arbitrarily call $t = 0$ "today" and $t = 1$ "tomorrow" (units are arbitrary). Future uncertainty is described by two possible states tomorrow, u and d, exactly one of which will be realized and observed tomorrow. To complete the model, we describe the probabilities of the states with a probability measure P on the "state space" $\Omega = \{u, d\}$, where $P(u) = p$ is the probability (as of today) that state u will occur, and correspondingly $P(d) = 1 - p$.

This is the simplest possible model of future uncertainty, with two uncertain future states and one time step. If $X : \Omega \to \mathbb{R}$ is a *claim* (random variable) representing a state-dependent payment realized tomorrow, what should we be willing to pay today to be guaranteed this payment?

In this book, we focus on arbitrage pricing, which means that we take the prices of some pre-specified securities to be given exogenously (as random variables, or random processes), and see what that implies for the possible prices of arbitrary claims.

In a nutshell, the idea of arbitrage pricing is that *if we can re-construct the claim X by a portfolio of existing securities, then the $t = 0$ price $V_0(X)$ of X must be the same as the price of the portfolio.* If not, we would take a short position in one and a long position in the other today, pocket the difference, and tomorrow owe nothing since the two positions would exactly cancel with certainty. This "arbitrage" strategy could be scaled to arbitrarily large size and therefore would crash the market unless prices adjusted to prevent this. The claim price therefore must be equal to the portfolio price, and it is totally unaffected by the preferences of agents or by the assumed probabilities of the future outcomes.

Let's see how this works. We introduce some securities: a bond B and a stock S, with positive prices and assume as usual that traders are allowed to hold any real valued number of shares of each, positive or negative.

The security prices are B_0 and S_0 at $t = 0$ and, at time $t = 1$, they are $B_1(u) = B_1(d) \equiv B_1$ (so that our bond is a deterministic, or riskless, security) and $S_1(u)$, $S_1(d)$. Without loss of generality, let's assume $S_1(d) < S_1(u)$, and note that a positive interest rate for the bond corresponds to the usual case $B_0 < B_1$. A portfolio π is a pair of real numbers $\pi = (\phi, \psi)$, where ϕ is the number of shares of the stock held from time 0 to time 1, and similarly ψ for the bond. The price of the portfolio at time t ($t = 0, 1$), is

$$V_t(\pi) = \phi S_t + \psi B_t, \tag{1.1}$$

so $V_1(\pi)$ is a random variable. The portfolio π is a *perfect hedge* for the claim X if $V_1(\pi) = X$ a.s. (that is, with probability one) as random variables.

For a given claim X, let us now find a perfect hedge. The condition is that ϕ and ψ should satisfy the (linear) equations

$$\phi S_1(u) + \psi B_1 = X(u) \tag{1.2}$$
$$\phi S_1(d) + \psi B_1 = X(d) \tag{1.3}$$

These are easily solved as

$$\phi = \frac{X(u) - X(d)}{S(u) - S(d)} \tag{1.4}$$
$$\psi = B_1^{-1}(X(u) - \phi S_1(u)) = B_1^{-1}(X(d) - \phi S_1(d)), \tag{1.5}$$

and so the resulting (unique) value of X at $t = 0$ is

$$V_0(X) = V_0(\pi) = \phi S_0 + \psi B_0$$

for these values of ϕ and ψ. After some algebraic rearranging, and introducing the *discount factor* $\beta = B_0 B_1^{-1}$, we have the explicit formula

$$V_0(X) = \beta \left[\left(\frac{\beta^{-1} S_0 - S_1(d)}{S_1(u) - S_1(d)} \right) X(u) + \left(\frac{S_1(u) - \beta^{-1} S_0}{S_1(u) - S_1(d)} \right) X(d) \right]. \tag{1.6}$$

Exercise 1.1 *Verify equation* (1.6).

It's tempting to write this in the form of an expectation:

$$V_0(X) = E_Q[\beta X], \tag{1.7}$$

where the expectation is taken with respect to the new purely formal probability measure Q defined by

$$Q(u) = \frac{\beta^{-1}S_0 - S_1(d)}{S_1(u) - S_1(d)} \text{ and } Q(d) = \frac{S_1(u) - \beta^{-1}S_0}{S_1(u) - S_1(d)}. \tag{1.8}$$

Note that $Q(u) + Q(d) = 1$; Q will be a probability measure provided both these values are non-negative, or

$$S_1(d) \leq \beta^{-1}S_0 \leq S_1(u). \tag{1.9}$$

This condition turns out to be financially natural. If it failed, say $\beta^{-1}S_0 < S_1(d)$, we then have

$$\frac{B_1}{B_0} < \frac{S_1(d)}{S_0} < \frac{S_1(u)}{S_0}. \tag{1.10}$$

This means that the return on the stock is certain to be greater than the return on the bond. If this is the case, there exists a simple arbitrage portfolio $\pi = (M/S_0, -M/B_0)$, for any $M > 0$, such that $V_0(\pi) = 0$ and $V_1(\pi) > 0$, giving us a certain payoff of arbitrarily large size for zero initial investment. Prohibiting this arbitrage amounts to requiring condition (1.9).

Exercise 1.2 *Fill in the details.*

The measure Q is called the pricing measure. It is unique, and we can summarize the above discussion by saying that the price of any claim is the expectation of the discounted claim value with respect to the pricing measure Q, defined from the asset prices in (1.8). We remark that Q has no evident relation to the measure P, the supposed actual probabilities of the states. At the moment, the formula $V_0(X) = E_Q[\beta X]$ is purely formal.

1.4   THE PRICING FORMULA

The pricing formula (1.6) gives the unique price of the arbitrary claim X forced by the condition of no arbitrage. Notice that this price does not depend on the preferences of the traders. Also, **the price does not depend on the original measure P**, the assumed probabilities of the two states. If we change the value of $P(u) = p$, this has no effect whatever on the price of X. This key point is worthy of some reflection. The price of the claim X depends only on the payoff values of X and the prices of the securities at $t = 1$ and $t = 0$.

Exercise 1.3 *As a sanity test, consider what happens in the limit as $p \to 1$. In this case, there is only one possible state u, so we now have a deterministic model where X has the sole value $X(u)$. Since $X(d)$ is now irrelevant, how could the formula (1.6) be true?* (Hint: consider what no arbitrage requires of the relationship between B_1 and $S(u)$ in this case.)

A brief look at the derivation of the pricing formula tells us that we found a unique solution for the price of X because we had two states and two securities, so that the resulting linear system was square. The unique solution exists because our assumption $S_1(d) < S_1(u)$ means the linear system is of full rank. This suggests how the story will go for more states and/or more securities.

1.5   RISKY BOND

We have so far taken the bond B to be deterministic for simplicity, but some reflection shows that this is not in any way necessary. Everything works out the same way with a stochastic bond $B_1(u) \neq B_1(d)$ (except the algebra takes a little more work), as we now describe.

The equations defining the hedging portfolio now become

$$\phi S(u) + \psi B(u) = X(u) \tag{1.11}$$
$$\phi S(d) + \psi B(d) = X(d) \tag{1.12}$$

where we have temporarily dropped the subscript "1" on S_1 and B_1 for convenience.

Assuming that the determinant $\Delta = S(u)B(d) - S(d)B(u)$ is nonzero, the unique solution is

$$\phi = \frac{B(d)X(u) - B(u)X(d)}{\Delta} \tag{1.13}$$
$$\psi = \frac{S(u)X(d) - S(d)X(u)}{\Delta}. \tag{1.14}$$

The claim value is again forced by the assumption of no arbitrage to be

$$\begin{aligned} V_0(X) &= \phi S_0 + \psi B_0 \\ &= \left(\frac{B_0 S(d) - B(d)S_0}{\Delta} \right) X(u) + \left(\frac{B(u)S_0 - S(u)B_0}{\Delta} \right) X(d). \end{aligned}$$

If we define the discount factor (now stochastic) as $\beta(\cdot) = B_0/B(\cdot)$, we obtain

$$V_0(X) = E_Q[\beta X], \tag{1.15}$$

where Q is a candidate probability measure defined by

$$Q(u) = \frac{-B(u)S(d) + B(u)B(d)(S_0/B_0)}{\Delta} \tag{1.16}$$
$$Q(d) = \frac{-B(d)B(u)(S_0/B_0) + S(u)B(d)}{\Delta}. \tag{1.17}$$

Exercise 1.4 *Verify formulas* (1.15), (1.16) *and* (1.17).

It is easy to see that $Q(u) + Q(d) = 1$. The following exercise verifies that Q is a true probability measure when there is no arbitrage in the market.

Exercise 1.5 *Show that Q is a probability measure as follows. For convenience and without loss of generality, assume that S and B are normalized so that $S_0 = 1 = B_0$ (in which case S_1 and B_1 represent returns). Show that if either $Q(u) < 0$ or $Q(d) < 0$, there is a simple arbitrage portfolio. (Hint: if $S(u) > B(u)$, argue that no arbitrage requires $S(d) < B(d)$.)*

The quantity S/B is called the *discounted stock price*. It will be useful later to notice that the pricing measure Q has a special property with respect to the discounted stock price: it makes the discounted stock a *martingale*, i.e.,

$$S_0/B_0 = E_Q[S_1/B_1]. \tag{1.18}$$

Exercise 1.6 *Verify this.*

1.6   GENERAL CASE OF ONE TIME STEP

Suppose now our finite state space is $\Omega = \{\omega_1, \omega_2, \ldots, \omega_m\}$, and we have $k+1$ securities $\{S^0, \ldots, S^k\}$, whose prices are known at time $t = 0$ but are functions of the state $\omega \in \Omega$ at $t = 1$. Let $X : \Omega \to \mathbb{R}$ be a claim paying off at time $t = 1$. By listing the values of X along the m states, we can think of X as a vector in $\mathbb{R}^m$.

A hedging portfolio for X is a vector $\phi = (\phi_0, \phi_1, \ldots, \phi_k) \in \mathbb{R}^{k+1}$ of holdings such that the portfolio determined by ϕ has value X at time $t = 1$ in every state. If we think of ϕ and X as column vectors, and define the $m \times (k+1)$ matrix $\mathcal{S}$ of asset prices at $t = 1$ by $\mathcal{S}(i, j) = S_1^j(\omega_i)$, $i = 1, \ldots, m$, $j = 0, \ldots, k$, then ϕ is determined by the $m \times (k+1)$ linear system of equations

$$\mathcal{S}\phi = X. \tag{1.19}$$

Suppose now that $m = k + 1$. If the square matrix $\mathcal{S}$ is invertible, the unique solution is given by $\phi = \mathcal{S}^{-1}X$; in other cases, there may be no solution or infinitely many solutions according to the usual story for systems of linear equations.

Assume for the remainder of this chapter that $\mathcal{S}$ is invertible so there is a unique solution for ϕ. Then the time-0 price of X is, assuming no arbitrage, given by the value of the hedging portfolio:

$$V_0(X) = \phi \cdot S_0 \tag{1.20}$$

where the dot indicates the usual scalar product. This can also be written

$$V_0(X) = \mathcal{S}^{-1}X \cdot S_0 = S_0' \mathcal{S}^{-1} X, \tag{1.21}$$

where in the last expression we are using matrix multiplication and the prime denotes transpose.

Formulas (1.20) and (1.21) depend on the prices of all the securities in all states, in general, but no one security is singled out yet. If we wish to write $V_0(X)$ as an expectation as we did before, we must select a security to use for the discount factor β – any of the securities may be used for this purpose. The chosen security is called the "numeraire", and can be thought of as a reference price against which to compare other prices. (In practice, this will usually be a money market account.)

By relabeling if needed, call the chosen numeraire S^0, so the discount factor is $\beta(\cdot) = S_0^0/S_1^0(\cdot)$, which is a random variable on Ω. Then exercise 1.7 below shows that our pricing formula is none other than the now-familiar equation

$$V_0(X) = E_Q[\beta X], \tag{1.22}$$

where the pricing measure Q is the unique probability measure making the discounted stock (vector) process a martingale:

$$S_0/S_0^0 = E_Q[S_1/S_1^0], \tag{1.23}$$

where $S_t/S_t^0 = (1, S_t^1/S_t^0, \ldots, S_t^k/S_t^0)$ for $t = 0, 1$.

Exercise 1.7 *Writing* $q_i = Q(\omega_i)$, *show that equation* (1.23) *is equivalent to the* $(k+1) \times (k+1)$ *system of linear equations*

$$\sum_{i=1}^{k+1} q_i \beta(\omega_i) S_1(\omega_i) = S_0. \tag{1.24}$$

Show that the solutions $\{q_i\}$ *satisfy* $\sum q_i = 1$, *and that*

$$E_Q[\beta X] \equiv \sum_{i=1}^{k+1} q_i \beta(\omega_i) X(\omega_i) = S_0' S^{-1} X = V_0(X). \tag{1.25}$$

What's missing from the story so far is why this Q must be a probability measure, i.e., $q_i \in [0, 1]$ for all i) in the absence of arbitrage; the proof of this is part of the Second Fundamental Theorem of Asset Pricing, which we discuss in Chapter 3.

CHAPTER 2

The General Discrete Model

We now move to the more interesting case of multiple time steps. Our view is that things work out too nicely for the usual binomial tree model to really see what's going on. So we're going to carry out the analysis in more generality: the general finite state, finite time model. The main ideas go back to Harrison and Pliska, though we follow our own tastes in this development.

(Readers should consult Appendix A: Probability Refresher for probability background and definitions of various terms used in this chapter, such as "filtration", "measurable", and so forth.)

2.1 THE TREE

We begin with a finite probability space (Ω, μ), the elements of which are called states. The measure μ is not really essential for us beyond the requirement that *every $\omega \in \Omega$ should have positive probability*. (If not, simply remove the zero probability elements.) We also want a finite set of times (at which decisions can be made), $t = 0, 1, 2, \ldots, T$, and a filtration $\{\mathcal{F}_t\} = \{\mathcal{F}_t : t = 0, \ldots, T\}$ on Ω such that $\mathcal{F}_0 = \{\emptyset, \Omega\}$ and $\mathcal{F}_T$ is the power set of Ω. Our model is then described by the triple $(\Omega, \mu, \{\mathcal{F}_t\})$.

For example, take any (non-recombining) tree with any finite branching you like at times $t = 0, 1, \ldots, T$. That is, start with a single node, or state, at time $t = 0$. From this "root node" a finite number of branches emanate, ending in nodes representing finitely many different states at time $t = 1$. Each of these nodes then branches finitely many times to reach nodes at time $t = 2$, etc. Every node has a unique path leading to it from the root node, but potentially many paths leading away from it toward the future.

The number of branches need not be the same from time to time or for different nodes at the same time. A choice of branch to follow at each time determines a *path* through the tree, representing a history of branch choices at each time.

We then consider each path starting at time 0 and ending at time T as a point ω in the sample space Ω, the set of all paths. A *probability measure on the tree* is just a probability measure on this set Ω, and we usually ignore paths with zero probability.

The tree structure determines a natural filtration, which is "history up to time t". The equivalence relation "ω_0 and ω_1 agree up to and including time t" gives equivalence classes that are the atoms of $\mathcal{F}_t$. Given a path ω, we denote by $\mathcal{F}_t[\omega]$ the atom of $\mathcal{F}_t$ containing ω, the set of paths agreeing with ω up to and including time t.

Exercise 2.1 *Given any finite model $(\Omega, \mu, \{\mathcal{F}_t\})$, construct a tree such that the space of paths of the tree, along with the natural filtration induced by the tree as above, is equivalent to $(\Omega, \mu, \{\mathcal{F}_t\})$.*

Exercise 2.1 shows that there is no loss of generality in describing general finite models in terms of trees, so we will do so whenever convenient.

Readers may be familiar with a common variant – a "recombining tree" which can have many paths to the same single node. Such trees can be obtained from our "non-recombining" trees by combining various sets of nodes at a common time into single nodes. A model based on such a recombining tree is equivalent to a model on a non-recombining tree in which the stock process (see below) is constrained to be constant on all the identified nodes. Therefore, without loss of generality we need not consider the recombining trees further.

2.2 THE STOCK PROCESS

Now consider also a stock process $S = \{S_t \text{ for } t = 0, \ldots, T\}$. Here S_t is a positive random variable on Ω which is $\mathcal{F}_t$-measurable, i.e., so that $S_t(\omega_0) = S_t(\omega_1)$ whenever ω_0 and ω_1 agree up to time t. This just means we decorate the nodes of the tree with positive stock prices.

Let $\mathbb{R}^+$ denote the positive real numbers. A convenient level of generality is to let the S_t's be $(\mathbb{R}^+)^{k+1}$-valued random variables. So now

$$S_t = (S_t^0, \ldots, S_t^k)$$

is a list of positive prices of $k + 1$ securities, and we agree that the first one, S_t^0, is our numeraire. For convenience, we'll call it the bond, but it need not be distinguished in any way from the other stocks, and in particular need not be riskless. S itself is called either the *stock process* or the *asset process*. Each component has units of dollars (\$). With the specification of the underlying stock process S, the asset pricing model $(\Omega, \mu, \{\mathcal{F}_t\}, S)$ is now completely specified.

2.3 TRADING STRATEGIES AND ATTAINABLE CLAIMS

A *trading strategy* ϕ is a predictable $\mathbb{R}^{k+1}$-valued process, i.e.,

$$\phi = \{\phi_t : t = 1 \ldots T\} = \{(\phi_t^0, \ldots, \phi_t^k) : t = 1 \ldots T\}$$

such that for $t = 1, \ldots, T$,

$$\phi_t \text{ is } \mathcal{F}_{t-1}\text{-measurable.}$$

The interpretation is that from time $t - 1$ to time t, ϕ_t is the dimensionless vector of holdings of the $k + 1$ securities. How exactly we choose to rebalance can depend on the entire history of security prices up to that point. It's important to get the subscripts straight — note that we start off with the portfolio ϕ_1 at time 0.

Having fixed a stock process earlier, associated to any trading strategy ϕ is another ($\mathbb{R}$-valued) process $V(\phi)$, the *value process*, giving the value of the indicated portfolio. For t between 1 and T, this is defined by

$$V_t(\phi) = \phi_t \cdot S_t$$

and for $t = 0$ we have $V_0 = \phi_1 \cdot S_0$. (The dot is the usual dot product in $\mathbb{R}^{k+1}$.) Note that V_t is $\mathcal{F}_t$-measurable since S_t is, so that at time t we find out what the portfolio is worth that was selected at time $t - 1$. The units of $V_t(\phi)$ are dollars at time t, just like S_t.

The only sort of strategy we care about is the self-financing kind. This means that at time $t - 1$ we are holding ϕ_{t-1}, we want to rebalance to ϕ_t, and our purchases should be exactly financed by our sales. In other words, the rebalanced portfolio should have the same value as the original one, or

$$\phi_{t-1} \cdot S_{t-1} = \phi_t \cdot S_{t-1}. \tag{2.1}$$

A predictable $k + 1$-dimensional process ϕ satisfying condition (2.1) for each $t = 1, \ldots, T$ is called *self-financing with respect to S*.

This is worth a few more words. For a general trading strategy, self-financing or not, the change in value of the indicated portfolio is

$$\begin{aligned} \Delta V_t &= V_t - V_{t-1} \\ &= \phi_t \cdot S_t - \phi_t \cdot S_{t-1} + \phi_t \cdot S_{t-1} - \phi_{t-1} \cdot S_{t-1} \\ &= \phi_t \cdot \Delta S_t + S_t \cdot \Delta \phi_t - \Delta S_t \Delta \phi_t \end{aligned}$$

which is just a discrete version of the product rule for differentiation.

If we write this instead as

$$\Delta V_t = \phi_t \cdot \Delta S_t + S_{t-1} \cdot \Delta \phi_t,$$

the first term is the change in the value of the portfolio due to changes in security prices, and the second is the change due to rebalancing alone. The self-financing condition is precisely that the second term vanishes.

Example. Consider the following strategy: do nothing until time t_0, then borrow \$1 of the numeraire, buy \$1 of stock j, and hold until time T. We name this strategy

$$\Phi^{t_0,j} = (\phi_1, \ldots, \phi_T)$$

where $\phi_1 = 0 = \phi_2 = \cdots = \phi_{t_0}$,

$$\phi_{t_0+1} = (-1/S_{t_0}^0, 0, \ldots, 0, 1/S_{t_0}^j, 0, \ldots, 0),$$

(where the second nonzero coordinate is in the jth position, counting from zero), and $\phi_{t_0+k} = \phi_{t_0+1}$ for $k = 2, \ldots, T - t_0$.

This strategy is self-financing. To verify this, we check $\Delta \phi_t \cdot S_{t-1} = 0$ for all t. Since ϕ is constant in t except at $t = t_0 + 1$, we only have to check that $\Delta \phi_{t_0+1} \cdot S_{t_0} = 0$, which follows easily from the definition of $\Phi^{t_0,j}$.

We note for future reference that

$$V_0(\Phi^{t_0,j}) = 0 \quad \text{and} \quad V_T(\Phi^{t_0,j}) = -\frac{S_T^0}{S_{t_0}^0} + \frac{S_T^j}{S_{t_0}^j}.$$

Exercise 2.2 *For an event $A \in \mathcal{F}_{t_0}$, define the strategy $\Phi_A^{t_0,j}$ by setting $\Phi_A^{t_0,j}(\omega) = \Phi^{t_0,j}(\omega)$ if $\omega \in A$, and otherwise zero. That is, we do nothing until time t_0, at which time we check to see whether the event A has occurred. If it has, we short a dollar of the numeraire to buy a dollar of stock j as before; if it hasn't, we do nothing.*

Show that $\Phi_A^{t_0,j}$ is self-financing,

$$V_0(\Phi_A^{t_0,j}) = 0 \text{ and } V_T(\Phi_A^{t_0,j}) = (-\frac{S_T^0}{S_{t_0}^0} + \frac{S_T^j}{S_{t_0}^j})\mathbb{1}_A. \tag{2.2}$$

A *claim* is simply a (real-valued) random variable X defined on Ω. Think of the claim X as a contractually agreed payoff of $X(\omega)$ dollars at time T if stock prices follow the particular path ω. Note that claims may take positive or negative values, or both (e.g., a forward contract). Therefore, any financial derivative paying off at the terminal time T is represented by a claim in our sense. The set of all claims is $\mathbb{R}^\Omega$, the set of all real-valued functions on Ω.

(One kind of derivative not covered by this definition is that of an American option because the holder may decide to exercise prior to the maturity time T. However, if we consider an American option along with an explicit exercise strategy that invests any early-exercise proceeds in a portfolio of the market securities, then this also becomes a claim in the above sense.)

A claim is *attainable* if there is a self-financing trading strategy which replicates it, in other words if there is a self-financing strategy ϕ such that $V_T(\phi) = X$. The attainable claims are the ones that have "arbitrage enforced prices", as we describe in the next section. But first, we further describe the set of attainable claims and the set of self-financing strategies.

2.4 ARBITRAGE

Loosely speaking, an arbitrage is a way of making a riskless profit. One must be very careful in making this statement precise since there is in fact a subtle gradation of types of arbitrage.

Definition 2.1 Arbitrage. Let $(\Omega, \mu, \{\mathcal{F}_t\}, S)$ be a finite asset pricing model as described above.

1. A *(weak) arbitrage* is a self-financing strategy ϕ with $V_0(\phi) = 0$ and $V_T(\phi) \geq 0$ and $\mu(V_T(\phi) > 0) > 0$.

2. A *strong arbitrage* is a self-financing strategy ϕ with $V_0(\phi) = 0$ and $V_T(\phi) > 0$.

3. An *inconsistent pricing strategy* is a self-financing strategy ϕ with $V_T(\phi) \equiv 0$ and $V_0(\phi) < 0$.

A market is *viable* if it admits no (weak) arbitrage.

For our purposes, the most important condition is the first one, the weak arbitrage. Unless otherwise noted, "arbitrage" always means "weak arbitrage". Think of it as a free lottery ticket — it costs nothing to set up, will not lose money, and might make money. Note that only the equivalence class of the measure μ has made an appearance here, in the form of asserting that a certain set has positive measure. Changing to a different equivalent measure (that is, having the same sets of measure zero) would not change the collection of arbitrage strategies.

By contrast, a strong arbitrage is a free lunch — it costs nothing to set up and will definitely make money (with probability one). Evidently, a strong arbitrage is an arbitrage. Some texts refer to the equivalent condition of the existence of a *dominant strategy*: a self-financing strategy ϕ is a dominant strategy if there is a self-financing strategy ψ such that $V_0(\phi) = V_0(\psi)$ and $V_T(\phi) > V_T(\psi)$.

For discrete models, the existence of a strong arbitrage is also equivalent to the existence of a self-financing strategy ϕ such that $V_0(\phi) < 0$ and $V_T(\phi) \geq 0$. Some texts define strong arbitrage this way.

Exercise 2.3 *Prove the equivalence asserted in the previous paragraph.*

An inconsistent pricing strategy is the most extreme case. While a strong arbitrage is a free guarantee today of a positive payment at maturity, an inconsistent pricing strategy is free money today, with no later obligations – there's no requirement to wait. It implies the existence of a strong arbitrage since if ϕ is an inconsistent pricing strategy, one may modify the strategy by adding to the portfolio $-V_0(\phi)/S_0^0$ shares of S_0, held until maturity.

A related concept is the following:

Definition 2.2   A market satisfies the *Law of One Price* if every two self-financing strategies that replicate the same claim have the same initial value.

Exercise 2.4 *Prove the following are equivalent:*

1. *there exists an inconsistent pricing strategy*

2. *for any attainable claim X and any initial price p, there is a self-financing strategy ϕ with $V_0(\phi) = p$ and $V_T(\phi) = X$*

3. *the Law of One Price fails.*

The relations among these types of arbitrage described above may be summarized as follows.

Proposition 2.3   *A viable market does not admit strong arbitrage. A market that does not admit strong arbitrage also does not admit inconsistent pricing.*

Exercise 2.5 *Prove Proposition 2.3*

This means that in a viable market none of these kinds of arbitrage exist. It follows easily from the absence of inconsistent pricing that any attainable claim must have a unique price at time zero (the Law of One Price), namely $V_0(\phi)$, the cost of setting up the replicating portfolio. This is sometimes called the "arbitrage enforced price" of the claim.

Another way to say this is that the following conditions are increasingly stronger assumptions about a market:

1. the Law of One Price holds

2. there is no strong arbitrage

3. the market is viable

Exercise 2.6 *Show that the above three conditions are distinct by finding examples of (a) a market for which the Law of One Price holds (there is no inconsistent pricing strategy), but there is a strong arbitrage, and (b) a market for which there is no strong arbitrage, but there is a weak arbitrage.*

The existence of arbitrage is a local property, in the sense that if there is an arbitrage strategy over the full time horizon from 0 to T, then in fact one can find a single state at a certain time such that an arbitrage exists in the next time step.

To be more precise, given $\omega_0 \in \Omega$, let $\mathcal{F}_t[\omega_0]$ denote the atom of $\mathcal{F}_t$ containing ω_0 – i.e., the set of paths that agree with ω_0 up to time t.

Definition 2.4   A *local arbitrage* is a self-financing strategy ϕ such that there exists a time t and a state $\omega_0 \in \Omega$ with $V_t(\phi)(\omega_0) = 0$, $V_{t+1}(\phi)(\omega) \geq 0$ for all $\omega \in \mathcal{F}_t[\omega_0]$, and $V_{t+1}(\phi)(\omega) > 0$ for at least one $\omega \in \mathcal{F}_t[\omega_0]$.

Equivalently, a market admits a local arbitrage if there exists a time t, a path $\omega_0 \in \Omega$, and a holdings vector h such that $h \cdot S_t(\omega_0) = 0$, $h \cdot S_{t+1}(\omega) \geq 0$ for all $\omega \in \mathcal{F}_t[\omega_0]$, and $h \cdot S_{t+1}(\omega) > 0$ for at least one $\omega \in \mathcal{F}_t[\omega_0]$.

The market admits a local arbitrage if there is some time t and state such that a portfolio with zero time-t value can be constructed that is guaranteed to be non-negative at time $t + 1$ and has a positive probability of being positive at $t + 1$.

Proposition 2.5   *A market admits arbitrage if and only if it admits a local arbitrage.*

As a practical matter, this means that if one is presented with a tree and security prices on it, one may determine whether an arbitrage exists by inspecting each node individually for a local arbitrage.

Proof: First suppose there is a local arbitrage at time t. To create an arbitrage strategy, first do nothing until time t. Then, if state ω_0 obtains, trade into holding vector h (at zero cost). At time $t + 1$, cash out any positive holdings into the numeraire and hold. The result at time T will be nonnegative with at least one positive state.

Conversely, suppose there exists an arbitrage strategy ϕ. Consider the set Λ of times $s \leq T$ such that $V_s(\phi)$ is nonnegative for all states and positive in at least one state. Then $T \in \Lambda$, so $\Lambda \neq \emptyset$. Let t be the least element of Λ. Note $t > 0$, because an arbitrage by definition has value 0 at $t = 0$.

We now argue that at $t - 1$ there is a local arbitrage. By choice of t, there are only two possibilities: either $V_{t-1}(\phi)$ is negative in some state $\omega_0 \in \Omega$, or else it is zero in all states.

If the former, then ω_0 is the desired state since one needs only to add a positive position in the numeraire to bring the value in that state up to zero and thereby also making the value at time t positive.

If the latter, then choose a time $t - 1$ state which leads to one of the positive values of $V_t(\phi)$. $\square$

For strong arbitrage and inconsistent pricing, local and global are not equivalent because global implies but is not implied by local.

Exercise 2.7 *Prove: A market admits strong arbitrage (respectively, inconsistent pricing) only if it admits a local strong arbitrage (respectively, local inconsistent pricing), but not conversely.*

Since global still implies local, one may verify that there is no strong arbitrage (or inconsistent pricing) in a given market model by examining each node of the tree separately and determining there is no local strong arbitrage (respectively, local inconsistent pricing) at any node.

In the next chapter, we find that determining the absence of arbitrage in a market model implies the existence of a pricing measure. Three fundamental theorems will tell the story.

CHAPTER 3

The Fundamental Theorems of Asset Pricing

In this chapter, we illuminate the importance of martingale measures, and their relation to asset pricing, the no arbitrage condition, and to market completeness. We use the definitions from the previous chapter. Recall the definition of martingale:

Definition 3.1 If $(\Omega, \mu, \{\mathcal{F}_t\})$ is a filtered measure space and $Y = \{Y_t : t = 0, \ldots, T\}$ is an $\{\mathcal{F}_t\}$-adapted random process, we say that Y *is a μ-martingale*, if for all s, t with $0 \leq s \leq t \leq T$,

$$Y_s = E_\mu(Y_t | \mathcal{F}_s).$$

A vector-valued process is a μ-martingale if each component is.

Now suppose we are given an asset pricing model $(\Omega, \mu, \{\mathcal{F}_t\}, S)$ as described in the previous chapter, where S is an $(\mathbb{R}^+)^{k+1}$-valued stock price process with first component S^0. Our discussion is best separated into three parts. The first explains why we are interested in martingales at all.

Theorem 3.2 First Fundamental Theorem of Asset Pricing. *Suppose v is any measure such that S/S^0 is a v-martingale. For an attainable claim X with replicating strategy ϕ and $0 \leq t \leq T$, we have*

$$V_t(\phi) = E_v\left(X \frac{S_t^0}{S_T^0} \middle| \mathcal{F}_t\right).$$

(Note that the following proof does not require v equivalent to μ.) We give two proofs.

First Proof: We simply calculate the expectation conditionally. To begin with, $E_v\left(X \dfrac{S_t^0}{S_T^0} \middle| \mathcal{F}_t\right) = E_v\left(V_T(\phi) \dfrac{S_t^0}{S_T^0} \middle| \mathcal{F}_t\right)$. Now for $\tau > t$ we have

$$
\begin{aligned}
E_v\left(V_\tau(\phi)\frac{S_t^0}{S_\tau^0}\bigg|\mathcal{F}_t\right) &= E\left(E\left(V_\tau(\phi)\frac{S_t^0}{S_\tau^0}\bigg|\mathcal{F}_{\tau-1}\right)\bigg|\mathcal{F}_t\right) \\
&= E\left(E\left(\phi_\tau\cdot S_\tau\frac{S_t^0}{S_\tau^0}\bigg|\mathcal{F}_{\tau-1}\right)\bigg|\mathcal{F}_t\right) \\
&= E\left(S_t^0\phi_\tau\cdot E\left(\frac{S_\tau}{S_\tau^0}\bigg|\mathcal{F}_{\tau-1}\right)\bigg|\mathcal{F}_t\right) \\
&= E\left(S_t^0\phi_\tau\cdot \frac{S_{\tau-1}}{S_{\tau-1}^0}\bigg|\mathcal{F}_t\right) \\
&= E\left(\phi_{\tau-1}\cdot S_{\tau-1}\frac{S_t^0}{S_{\tau-1}^0}\bigg|\mathcal{F}_t\right) \\
&= E\left(V_{\tau-1}(\phi)\frac{S_t^0}{S_{\tau-1}^0}\bigg|\mathcal{F}_t\right)
\end{aligned}
$$

by properties of conditional expectation, predictability of ϕ, definition of martingale, and self-financing of ϕ, in that order. Proceeding iteratively from $\tau = T$ to $\tau = t+1$, we find

$$
E\left(X\frac{S_t^0}{S_T^0}\bigg|\mathcal{F}_t\right) = E\left(V_t(\phi)\frac{S_t^0}{S_t^0}\bigg|\mathcal{F}_t\right) = V_t(\phi). \quad\square
$$

Second Proof: Consider the discounted value process

$$
\tilde{V}_t = \frac{V_t(\phi)}{S_t^0} = \phi_t\cdot\tilde{S}_t
$$

written here in terms of the discounted stock process $\tilde{S}_t = S_t/S_t^0$. The self-financing property of ϕ means $S_t\cdot\Delta\phi_{t+1} = 0$, so we also have $\tilde{S}_t\cdot\Delta\phi_{t+1} = 0$. For the discounted value process, this means

$$
\Delta\tilde{V}_{t+1} = \tilde{V}_{t+1} - \tilde{V}_t = \phi_{t+1}\cdot\Delta\tilde{S}_{t+1} + \tilde{S}_t\cdot\Delta\phi_{t+1} = \phi_{t+1}\cdot\Delta\tilde{S}_{t+1}.
$$

Predictability of ϕ immediately yields

$$
E\left(\Delta\tilde{V}_{t+1}\bigg|\mathcal{F}_t\right) = \phi_{t+1}\cdot E\left(\Delta\tilde{S}_{t+1}\bigg|\mathcal{F}_t\right) = 0.
$$

This says that $\tilde{V}_t$ is a martingale, so in particular

$$
\frac{V_t(\phi)}{S_t^0} = E\left(\frac{V_T(\phi)}{S_T^0}\bigg|\mathcal{F}_t\right).
$$

$\square$

We remark that the second proof above also establishes in passing the following useful fact:

Proposition 3.3 *A strategy is self-financing with respect to the stock process if and only if it is self-financing with respect to the discounted stock process.*

From Theorem 3.2, we may conclude that in a viable market the unique price of any attainable claim is equal to the expected value of the discounted claim using any martingale measure (that is, a measure making S/S^0 a martingale). In particular,

Corollary 3.4

1. *All martingale measures price the attainable claims equally, and*

2. *if there is a martingale measure, then all replicating strategies for a given claim have the same value at all times.*

Exercise 3.1 *Prove the above Corollary.*

Of course, we do not yet know whether any martingale measures exist!

The next theorem tells us what it means to say that our market is viable in terms of the existence martingale measures. To keep the proof concise, we will postpone commentary and narration until the next section on forwards.

Theorem 3.5   Second Fundamental Theorem of Asset Pricing.

Let $(\Omega, \mu, \{\mathcal{F}_t\}, S)$ be a discrete asset pricing model. Then

1. *there exists a probability measure ν equivalent to μ such that S/S^0 is a ν-martingale if and only if there are no arbitrage opportunities, and*

2. *there exists a probability measure ν, absolutely continuous with respect to μ but possibly inequivalent to μ, such that S/S^0 is a ν-martingale if and only if there are no strong arbitrage opportunities.*

Proof: One direction is easy. Suppose ν is a probability measure such that S/S^0 is a ν-martingale. Let ϕ be any self-financing strategy with $V_T(\phi) > 0$. Then $E_\nu(V_T(\phi)/S_T^0) > 0$, so by Theorem 3.2 $V_0(\phi) > 0$. Hence, there can be no strong arbitrage.

If ν is equivalent to μ, and ϕ is a self-financing strategy such that $V_T(\phi) \geq 0$ and $\mu(V_T(\phi) > 0) > 0$, then again $E_\nu(V_T(\phi)/S_T^0) > 0$, so again by Theorem 3.2 $V_0(\phi) > 0$. Hence, there can be no arbitrage.

The other direction is harder. In each case we must prove the existence of an appropriate martingale measure.

First, note that the space of claims is just the finite-dimensional Euclidean space $\mathbb{R}^\Omega$, i.e., a claim is a list of payoffs, one per path. We make the following definitions:

$$
\begin{aligned}
\mathcal{X}^+ &= \{X \in \mathbb{R}^\Omega : X \geq 0 \text{ and } \mu(X > 0) > 0\} \\
\mathcal{X}^{++} &= \{X \in \mathbb{R}^\Omega : X > 0\} \\
\mathcal{X}^0 &= \{X \in \mathbb{R}^\Omega : X = V_T(\phi) \text{ for some } \phi \text{ self-financing with } V_0(\phi) = 0\}
\end{aligned}
$$

We take up the proofs of sufficiency one at a time.

(1) The definition of no arbitrage is that $\mathcal{X}^+$ and $\mathcal{X}^0$ are disjoint. Note that $\mathcal{X}^+$ is exactly the closed positive orthant of $\mathbb{R}^\Omega$ minus the origin, and that $\mathcal{X}^0$ is a closed linear subspace. Hence, we may apply Proposition B.1 (proved in Appendix B without reference to any earlier material) to conclude that there exists a vector $(\lambda_i) \in \mathcal{X}^{++}$ orthogonal to $\mathcal{X}^0$. This defines a linear functional λ on $\mathbb{R}^\Omega$ by $\lambda(X) = \sum \lambda_i X(\omega_i)$ that is zero on $\mathcal{X}^0$ and positive on $\mathcal{X}^+$. We may normalize so that $\lambda(\mathbb{1}) = 1$.

Such a linear functional uniquely determines a measure, which we denote by the same symbol λ, by means of the formula

$$\lambda(X) = \int_\Omega X \, d\lambda = \sum X(\omega_i)\lambda_i.$$

Positivity of (λ_i) means that, as a measure, λ is positive on each path in Ω and hence is equivalent to μ.

Recall now our long-short self-financing strategy $\Phi_A^{t,n}$ defined in Exercise 2.2 for any time t, asset n, and event $A \in \mathcal{F}_t$. Let $X = V_T(\Phi_A^{t,n})$. From equations (2.2), we have

$$V_0(\Phi_A^{t,n}) = 0 \text{ and } V_T(\Phi_A^{t,n}) = \left(-\frac{S_T^0}{S_t^0} + \frac{S_T^n}{S_t^n} \right) \mathbb{1}_A.$$

Hence, $X \in \mathcal{X}^0$ so $\lambda(X) = 0$, i.e.,

$$\lambda\left(\left(-\frac{S_T^0}{S_t^0} + \frac{S_T^n}{S_t^n} \right) \mathbb{1}_A \right) = 0 \tag{3.1}$$

for any time t, asset n, and for all events $A \in \mathcal{F}_t$.

From the definition of conditional expectation, we therefore have

$$E_\lambda\left[\left(-\frac{S_T^0}{S_t^0} + \frac{S_T^n}{S_t^n} \right) | \mathcal{F}_t \right] = 0$$

or equivalently

$$E_\lambda\left[\frac{S_T^0}{S_t^0} | \mathcal{F}_t \right] = E_\lambda\left[\frac{S_T^n}{S_t^n} | \mathcal{F}_t \right] \tag{3.2}$$

for all t, n. Equation (3.2) means that the *λ-expected returns of all assets are equal, over all periods ending at T*. This is equivalent to

$$E_\lambda[S_T^n | \mathcal{F}_t] = E_\lambda\left[\frac{S_t^n}{S_t^0} S_T^0 | \mathcal{F}_t \right],$$

or

$$\lambda\left(\left(S_T^n - \frac{S_t^n}{S_t^0} S_T^0 \right) \mathbb{1}_A \right) = 0 \tag{3.3}$$

for all events $A \in \mathcal{F}_t$.

Notice that up to this point, we have made no use of the choice of S^0 as numeraire. Now that selection becomes important as we define ν by means of the Radon-Nikodym derivative

$$\frac{d\nu}{d\lambda} = \frac{S_T^0}{\lambda(S_T^0)}.$$

(The denominator $\lambda(S_T^0)$ is simply a normalization factor to ensure that ν has total mass 1.) That is, for any claim X,

$$\nu(X) = \sum X(\omega_i)\lambda_i \frac{S_T^0(\omega_i)}{\lambda(S_T^0)}.$$

Verifying that ν is a martingale measure is now easy. For any n and t, we need to establish that

$$\frac{S_t^n}{S_t^0} = E_\nu[\frac{S_T^n}{S_T^0}|\mathcal{F}_t].$$

This is equivalent to

$$\nu((\frac{S_T^n}{S_T^0} - \frac{S_t^n}{S_t^0})|\mathbb{1}_A) = 0 \text{ for all } A \in \mathcal{F}_t,$$

or, in terms of λ,

$$\lambda(\frac{1}{\lambda(S_T^0)}(S_T^n - \frac{S_t^n}{S_t^0}S_T^0)|\mathbb{1}_A) = 0,$$

i.e.,

$$\lambda((S_T^n - \frac{S_t^n}{S_t^0}S_T^0)|\mathbb{1}_A) = 0, \text{ for all } A \in \mathcal{F}_t.$$

This is equation (3.3), so we are done.

(2) The definition of no strong arbitrage is that $\mathcal{X}^{++}$ is disjoint from $\mathcal{X}^0$. By Proposition B.1, there is a vector (λ_i) orthogonal to $\mathcal{X}^0$ and contained in $\mathcal{X}^+$, that is, is non-negative with possibly some zero entries. This means the corresponding measure λ is non-negative but possibly inequivalent to μ. The remainder of the argument proceeds exactly as before.
□

Evidently, the unsung hero of the proof is the measure λ, which does not depend on a choice of numeraire, rather than the more popular ν. We know that ν-expectation of a discounted (with respect to S^0) claim gives the present value of the claim, but what can we say about λ? We will show that λ provides *forward prices* of claims consistent with no arbitrage. We will call λ the *forward measure* and provide more discussion in the next chapter.

The last part of the big theorem addresses the two sources of ambiguity in our pricing machinery, namely that our measure may not be unique, and we can only price attainable claims.

Definition 3.6 We say a market is *complete* if every claim is attainable.

Theorem 3.7   Third Fundamental Theorem of Asset Pricing.   *Assume the market* $(\Omega, \mu, \{\mathcal{F}_t\}, S)$ *admits no arbitrage. Then there exists exactly one measure* ν *equivalent to* μ *such that* S/S^0 *is a* ν-*martingale if and only if the market is complete.*

Proof: Let $\mathcal{A}$ denote the set of all attainable claims, a linear subspace of $\mathbb{R}^\Omega$.

Suppose not every claim is attainable. Then $\dim \mathcal{A} < |\Omega|$. (Note this is a finite dimensional proof.) Since $\dim \mathcal{X}^0 < \dim \mathcal{A}$, the space $\mathcal{X}^0$ must have codimension greater than one. By the proof of the previous theorem, this means λ, and hence ν is not unique.

Exercise 3.2 *Argue that uniqueness of* ν *implies uniqueness of* λ, *where* ν *and* λ *are the measures arising in the proof of Theorem 3.5.*

Conversely, suppose every claim is attainable, so $\mathcal{A} = \mathbb{R}^\Omega$. Let ν_1, ν_2 be two measures with respect to which S/S^0 is a martingale, and let X be any claim. Since $X \in \mathcal{A}$, X has a replicating strategy ϕ. By Theorem 3.2,

$$V_t(\phi) = S_t^0 E_{\nu_1}\left(\frac{X}{S_T^0}|\mathcal{F}_t\right) = S_t^0 E_{\nu_2}\left(\frac{X}{S_T^0}|\mathcal{F}_t\right)$$

for all t. Setting $t = 0$, we deduce that

$$E_{\nu_1}(Y) = E_{\nu_2}(Y)$$

for all claims Y. Hence, $\nu_1 = \nu_2$. $\square$

It is quite possible for all claims to be attainable even though there is no martingale measure at all. In this case, however, there must be arbitrage opportunities by Theorem 3.5.

Exercise 3.3 *Construct an example of a market for which all claims are attainable, but there are no martingale measures for the discounted stock process.*

CHAPTER 4

Forwards and Futures

4.1 FORWARDS AND THE FORWARD MEASURE

Suppose we have a constant continuously compounded interest rate r (bank account) and a forward contract on a single non-dividend-paying stock S_t, maturing at time T. That is, at time 0, we have agreed to pay a strike price K at time $T > 0$ and, in exchange, receive at that time one share of stock, with value S_T.

By the easy arbitrage argument of Section 1.2, in order that this contract have value zero at time 0, the value of K must be

$$K = S_0 e^{rT}.$$

This strike price for which the forward contract has initial value zero is called the "forward price" of the stock.

What if the interest rate is not constant or not deterministic? We next generalize this discussion in context of our discrete models; our main tool is the forward measure λ, first seen in the previous chapter.

Our setting again is a discrete market model $(\Omega, \mu, \{\mathcal{F}_t\}, S)$ with assets $S = (S^0, \ldots, S^k)$, an $\{\mathcal{F}_t\}$-adapted process in $(\mathbb{R}^+)^{k+1}$. We assume the market admits no arbitrage, and therefore by Theorem 3.5 and its proof, we have a martingale measure ν and a measure λ related by

$$\frac{d\nu}{d\lambda} = \frac{S_T^0}{\lambda(S_T^0)} \ .$$

We call λ a "forward measure" for reasons that will become clear shortly.

Exercise 4.1 *Verify that*

$$\frac{d\lambda}{d\nu} = \frac{1/S_T^0}{\nu(1/S_T^0)}.$$

The numeraire S^0 is now arbitrary. However, we note that if S^0 is taken to be deterministic, or even merely that S_T^0 is deterministic (so that S^0 is a bond maturing at T), then

$$\frac{d\nu}{d\lambda} = \frac{S_T^0}{\lambda(S_T^0)} = 1$$

so in this case $\lambda = \nu$.

The forward measure λ can be used as an alternative to the martingale measure ν to value attainable claims. If ϕ is a self-financing strategy replicating the claim X, then, using the change of measure formula for conditional expectation

$$E_\nu(X|\mathcal{F}_t) = \frac{E_\lambda\left(X\frac{d\nu}{d\lambda}\,|\,\mathcal{F}_t\right)}{E_\lambda(\frac{d\nu}{d\lambda}|\mathcal{F}_t)}$$

and Theorem 3.2,

$$\begin{aligned} V_t(\phi) &= E_\nu[X\frac{S_t^0}{S_T^0}|\mathcal{F}_t] \\ &= E_\lambda[X\frac{S_t^0}{S_T^0}\frac{d\nu}{d\lambda}|\mathcal{F}_t]/E_\lambda[\frac{d\nu}{d\lambda}|\mathcal{F}_t] \\ &= E_\lambda[XS_t^0|\mathcal{F}_t]/E_\lambda[S_T^0|\mathcal{F}_t] \end{aligned}$$

or

$$V_t(\phi) = d(t, T)E_\lambda[X|\mathcal{F}_t], \tag{4.1}$$

where

$$d(t, T) = \frac{1}{E_\lambda[\frac{S_T^0}{S_t^0}|\mathcal{F}_t]}$$

is an $\mathcal{F}_t$-measurable discount factor independent of the choice S^0 of numeraire by equation (3.2). Equation (4.1) is a valuation formula where the discount factor appears outside the expectation instead of inside. As we describe next, the term $E_\lambda[X|\mathcal{F}_t]$ turns out to be none other than the time t forward price of X for maturity T.

A claim $\mathbb{1}$ paying 1 at time T in all states of the world is called a "T-bond". The forward measure is sometimes described as the equivalent martingale measure we get when we choose the T-bond for our numeraire. However, this presumes that the T-bond is attainable. In an incomplete market where the T-bond is not attainable, we still have forward measures but they are not unique.

Since the forward measure depends on the time T to maturity, it is often called a "T-forward measure" λ_T.

Definition 4.1   Let X be a claim paying off at time T, t a time with $0 \le t < T$, and K an $\mathcal{F}_t$-measurable function. The *forward contract on X struck at K at time t and maturing at T* is the claim $X - K$.

The (t, T)-*forward price of X* is the strike K for which this forward contract has value 0 at time t.

Our interpretation is that the (t, T)-forward price is the fair price, agreed upon at time t, to exchange for delivery of X at time T. It is somewhat misleading to speak of "the" forward price of an attainable claim X unless the T-bond is also attainable, since otherwise the value of K for which $X - K \cdot \mathbb{1}$ has present value zero depends on the choice of ν, hence λ_T. Therefore, "the" is always taken to be relative to the choice of λ_T. We don't have to worry if both X and the T-bond are attainable, for example in a complete market.

The following theorem helps explain why we call λ_T a "forward measure".

Theorem 4.2 *In a viable market, the (t, T)-forward price of a claim X is $K = E_{\lambda_T}(X|\mathcal{F}_t)$. Moreover, if X and $\mathbb{1}$ are attainable, then K is independent of the choice of λ_T.*

Proof: Via the statement and proof of Theorem 3.5, consider a measure ν making S/S^0 a ν-martingale and the corresponding λ_T, determining ν by $\frac{d\nu}{d\lambda_T} = \frac{S_T^0}{\lambda_T(S_T^0)}$. To ease notation denote λ_T by λ. For a claim X, the forward price K should satisfy

$$
\begin{aligned}
0 &= E_\nu\left((X - K)\frac{S_t^0}{S_T^0} \Big| \mathcal{F}_t \right) \\
&= \frac{E_\lambda\left(S_t^0(X - K) \big| \mathcal{F}_t \right)}{E_\lambda(S_T^0|\mathcal{F}_t)} \\
&= \frac{S_t^0\left(E_\lambda(X|\mathcal{F}_t) - K \right)}{E_\lambda(S_T^0|\mathcal{F}_t)}
\end{aligned}
$$

and the theorem follows. We have again made use of the change of measure formula for conditional expectations.

For the uniqueness of K, since $d\lambda/d\nu = (1/S_T^0)(1/\nu(1/S_T^0))$ (see Exercise 4.1), we have

$$
\begin{aligned}
E_\lambda[X|\mathcal{F}_t] &= E_\nu[X\frac{d\lambda}{d\nu}|\mathcal{F}_t]/E_\nu[\frac{d\lambda}{d\nu}|\mathcal{F}_t] \\
&= E_\nu[X\frac{S_t^0}{S_T^0}|\mathcal{F}_t]/E_\nu[\mathbb{1}\frac{S_t^0}{S_T^0}|\mathcal{F}_t].
\end{aligned}
$$

The latter expression is the ratio of the time-t price of X to the time-t price of the T-bond; since X and $\mathbb{1}$ are attainable, their prices are independent of the choice of ν, hence of λ. $\square$

An immediate consequence is that forward prices are independent of the choice of numeraire (since λ is). How can this be compatible with the discussion that began this chapter? There, the formula $K = Se^{r(T-t)}$ seems to have the discounting (and so the bank account as numeraire) built inextricably into it. An explanation comes from our observation in equation (3.2) that $E_\lambda(\frac{S_T^k}{S_t^k}|\mathcal{F}_t)$ is independent of k. Calling this expectation $1/d(t, T)$, we find that the time t forward price of S^k is

$$
E_\lambda(S_T^k|\mathcal{F}_t) = S_t^k/d(t, T),
$$

i.e., spot over the same discount factor that applies to all securities in the given state of the world. In the event that one of the securities is a riskless bond, i.e., S_t^0 is constant for each t), then this discount factor would be the usual S_t^0 / S_T^0, but, in general, $d(t, T)$ varies from node to node at time t.

The useful fact about λ that all securities have the same forward return has a simple arbitrage explanation. If S^0 had lower forward return than S^1, an arbitrageur would go long the forward contract on S^0, short \$1 of S_0, short the forward contract on S^1, and long \$1 of S^1. Then at time T he delivers the S^1 at the agreed-upon higher return (times \$1) than he pays for the S^0 he has agreed to buy (and return from the short sale at time 0).

Earlier we remarked that when the numeraire is deterministic, the forward measure λ is equal to the martingale measure ν. This means that the common assumption of a constant interest rate in treatments of asset pricing tends to obscure the distinction between these two measures, and it may be partly why the very valuable forward measure gets less attention than it deserves.

4.2   FUTURES

Having just described the importance and numeraire independence of forward contracts, we turn to contrast with the nature of futures contracts. This topic is different from ones handled so far, not least because a futures contract entails cash flows not just at the horizon T, but at all intervening times as well.

First, we should define what we mean by futures. For forwards, we first defined "forward contract with strike K," then "forward price" as the strike giving present value zero to that contract. Futures work a bit differently.

Fix a claim X. Then the futures price process $F_t(X)$ is a certain sequence of random variables that we will determine more explicitly below. A futures contract on X may be entered into at any moment t at no charge and closed at any subsequent moment at no charge. At the end of each period during which the contract is open, say time $t + 1$, the long position sees a cash flow of $F_{t+1}(X) - F_t(X)$ dollars. (The contract is *marked to market* after each period by having this amount added to or deducted from the contract holder's margin account.) If the contract is still open at time T, the long position actually receives X, so we must have $F_T(X) = X$.

The defining property of the futures price F_t is that the expected value of discounted cash flows from the contract is zero.

The reader will note that we have not said anything about arbitrage. In fact, we have been less than forthright by choosing a numeraire ("...discounted cash flows ...") without mentioning it. The first surprise about futures prices is that $F_t(X)$ depends on the choice of numeraire (in stark contrast to forward prices). Moreover, the numeraire needs to be predictable. We enshrine this requirement in a definition.

Definition 4.3   A *money market account* is a security S whose value S_t is $\mathcal{F}_{t-1}$-measurable.

Note that a money market account still allows for random interest rates, as long as they are known at the beginning of the period.

Theorem 4.4   *Suppose S^0 is a money market account and v is a measure equivalent to μ such that $(S/S^0, v, \mathcal{F})$ is a martingale. Let X be a claim. Then the process $F_t = E_v[X|\mathcal{F}_t]$ is the unique $\mathcal{F}$-adapted process $F_t = F_t(X)$ such that*

1. $F_T = X$, and

2. *for all s, t such that $0 \le s < t \le T$, $E_v((F_t - F_{t-1})\frac{S_s^0}{S_t^0}|\mathcal{F}_s) = 0$.*

The second condition means that, at any time s, the v-value of each of the future cash flows of the contract is zero, and hence the whole contract has value zero at all times.

Proof: It is straightforward to check that the process $F_t = E_v[X|\mathcal{F}_t]$ satisfies conditions 1 and 2 of the theorem. Conversely, suppose F_t is an $\mathcal{F}$-adapted process satisfying conditions 1 and 2.

For any $t > 0$, since S_t^0 and S_{t-1}^0 are both $\mathcal{F}_{t-1}$ measurable, condition 2 implies

$$E_v[(F_t - F_{t-1})|\mathcal{F}_{t-1}] = 0.$$

Fixing an s, summing over $t > s$, and taking expectations conditional on $\mathcal{F}_s$ yields

$$\sum_{t=s+1}^{T} E_v[F_t - F_{t-1}|\mathcal{F}_s] = 0,$$

and since $F_T = X$ and this is a telescoping sum, we get

$$E_v[X - F_s|\mathcal{F}_s] = 0$$

or $F_s = E_v[X|\mathcal{F}_s]$ for all s as desired. $\square$

Definition 4.5   The process F_t of the previous theorem is called the *T-futures prices of X at time t relative to S^0 and v.*

Corollary 4.6   *If the numeraire is deterministic, then the futures price and the forward price are equal.*

Proof: As mentioned before, for a deterministic numeraire $dv/d\lambda = 1$, so $\lambda = v$. This means $E_v[X|\mathcal{F}_t] = E_\lambda[X|\mathcal{F}_t]$ for all X and t. $\square$

Comment: For our discrete time setting, it was crucial for the proof of the theorem that the numeraire be a money market account. From a real world standpoint, this is probably not such a bad thing. In continuous time, remarkably enough, the proof goes through for arbitrary numeraire (see [EK99], p. 216-7), essentially because the distinction between S_t^0 and S_{t+1}^0 obstructing the discrete version goes away.

4.3   THE CONVEXITY CORRECTION

We know that, in general, futures and forward prices disagree; their difference is informally known as a "convexity correction". Given a market-quoted futures price, how should this be adjusted to obtain the corresponding forward price (with the same maturity)?

Here we look at the *relative convexity correction* given by (futures price – forward price)/(futures price), which is the percent change to be subtracted from the futures price to arrive at the forward price. Recall that $E_\lambda[X]$ and $E_\nu[X]$ are the $t = 0$ forward and futures prices (respectively) of X (maturity T). For convenience, we use the earlier notation $\lambda(X)$ and $\nu(X)$ for these.

Theorem 4.7   *The relative convexity correction between time $t = 0$ futures and forward prices of a claim X, with maturity T, is*

$$\frac{\nu(X) - \lambda(X)}{\nu(X)} = \lambda\left(\frac{S_T^0 - \lambda(S_T^0)}{\lambda(S_T^0 X)} X\right).$$

Proof: Change the ν-expectations to λ-expectations using the Radon-Nikodym derivative $d\nu/d\lambda = S_T^0/\lambda(S_T^0)$; the resulting computation is straightforward. □

Exercise 4.2 *(a) Fill in the details of the proof above.*

(b) Show that the convexity correction is zero if any of the following conditions is true:

1. *S_T^0 is deterministic i.e., S^0 is a T-bond)*

2. *X is deterministic*

3. *X and S^0 are uncorrelated with respect to the measure λ, i.e., $\lambda(XS^0) = \lambda(X)\lambda(S^0)$.*

A typical application for the convexity correction is to interest rate assets like zero coupon bonds. In our case, if we look at contracts on the numeraire, $X = S_T^0$, we get a simplified formula:

$$
\begin{aligned}
\frac{\nu(S_T^0) - \lambda(S_T^0)}{\nu(S_T^0)} &= \lambda\left(\frac{S_T^0 - \lambda(S_T^0)}{\lambda((S_T^0)^2)} S_T^0\right) \\
&= \frac{\lambda((S_T^0)^2) - (\lambda(S_T^0))^2}{\lambda((S_T^0)^2)} = \frac{\operatorname{var}(S_T^0)}{\lambda((S_T^0)^2)}.
\end{aligned}
$$

This latter expression is reminiscent of the formula for the curvature of a convex function and gives a hint about the origin of the common usage "convexity". Notice that since this expression is non-negative, the futures price of the numeraire is never less than its forward price.

4.4   COMPUTATIONAL MATTERS

We now turn to some practical issues in calculating prices of claims, as well as their forward and futures prices, for discrete market models $(\Omega, \mu, \{\mathcal{F}_t\}, S)$, where S is the vector of $\{\mathcal{F}_t\}$-adapted stock price processes $S = (S^0, \ldots, S^k)$.

Recall that the measure μ does not enter into the pricing formulas, so our data is really just a finite time-stratified tree with a single node at $t = 0$, with each path branching an arbitrary finite amount at each time step until $t = T$, along with prices for the $k + 1$ stocks at each node. A claim X is merely a function defined on the set of all paths.

As we've shown previously, all pricing questions can be answered by finding a martingale measure ν (unique if the market is complete) and the associated forward measure λ. Recall that these are related by

$$\frac{d\nu}{d\lambda} = \frac{S_T^0}{\lambda(S_T^0)},$$

or, equivalently,

$$\nu(X) = \lambda(X \frac{S_T^0}{\lambda(S_T^0)}).$$

From Exercise 4.1 this is also equivalent to

$$\frac{d\lambda}{d\nu} = \frac{1/S_T^0}{\nu(1/S_T^0)},$$

or, equivalently,

$$\lambda(X) = \nu(X \frac{1/S_T^0}{\lambda(1/S_T^0)}).$$

It's normally simpler to calculate ν first, then obtain λ by the formula above.

By our fundamental theorems, the existence of ν follows from no arbitrage, which, we recall, is a local condition. Therefore, our computational strategy is to compute a local martingale measure at each node; the global measure of a path will then simply be the product of the local measures along the path.

This observation reduces the problem to solving systems of linear equations. For example, suppose we examine a node at time t from which emanate j branches. Locally, our requirement is that the process

$$S/S^0 = (1, S^1/S^0, \ldots, S^k/S^0) = (1, \tilde{S}^1, \ldots, \tilde{S}^k)$$

should be a martingale with respect to some local measure $q = (q_1, \ldots, q_j)$, where q_i is the local measure assigned to branch $i, i = 1, \ldots, j$. In other words, if we use the notation $S_{t+1}^n(i)$ to denote the price of security n at time $t + 1$ on the ith branch of our node, we need

$$\sum_{i=1}^{j} q_j(1, \tilde{S}_{t+1}^1(i), \ldots, \tilde{S}_{t+1}^k(i)) = (1, \tilde{S}_t^1, \ldots, \tilde{S}_t^k).$$

This is a linear system of $k + 1$ equations in the j unknowns $q_1, \ldots, q_j$. (Note the first equation simply says $\sum q_j = 1$.) If we assume our market is viable, this has at least one solution, and then, in general, there are the usual three cases:

1. if $j < k + 1$, then the market is overdetermined, and some assets are redundant at this node;

2. if $j = k + 1$, we get a unique local measure q at this node; if the number of branchings equals the number of assets at all nodes, there is then a unique martingale measure and the market is complete;

3. if $j > k + 1$, there are infinitely many solutions, hence infinitely many martingale measures and the market is incomplete.

We typically see cases 2 or 3; case 2 covers the familiar binomial tree model with two assets, a stock and an bond.

Example: Consider a two-time-step (non-recombining) binomial tree model where the nodes are designated $\{0, u, d, uu, ud, du, dd\}$.

Let the asset prices (B, S) be given as follows: $(B, S)_0 = (1, 1)$,

$$[(B, S)(u), (B, S)(d)] = [(1, 2), (1, 0.5)], \text{ and}$$

$$[(B, S)(uu), (B, S)(ud), (B, S)(du), (B, S)(dd)] = [(2, 6), (2, 2), (1, 1), (1, 0.25)].$$

Note B is previsible, so it makes sense to ask about futures prices using B as numeraire.

Solving for the local weights at each of the three nodes and multiplying along paths gives us the values of the unique martingale measure:

$$\nu(uu, ud, du, dd) = (1/6, 1/6, 2/9, 4/9).$$

Since $\nu(1/S_2^0) = 5/6$, we have $\lambda(X) = (6/5)\nu(X/S_2^0)$, so we obtain

$$\lambda(uu, ud, du, dd) = (1/10, 1/10, 4/15, 8/15).$$

Now we can compute prices using our valuation formulas, e.g., the $(0,2)$-forward price of the stock S is

$$E_\lambda[S_2] = \lambda(S_2) = (1/10) \cdot 6 + (1/10) \cdot 2 + (4/15) \cdot 1 + (8/15) \cdot (1/4) = 6/5.$$

The corresponding futures prices is

$$E_\nu[S_2] = (1/6) \cdot 6 + (1/6) \cdot 2 + (2/9) \cdot 1 + (4/9) \cdot (1/4) = 5/3.$$

Exercise 4.3 Trinomial tree. *Consider the two-time-step trinomial tree: one node 0 at time 0, three nodes u,m,d at time 1, and nine nodes uu, um, ud, mu, mm, md, du, dm, dd at time 2. There are three assets (S^0, S^1, S^2) with time zero price $(1, 1, 1)$, time 1 prices $(1,2,1)$, $(1,1,2)$, and $(1, 1/2, 1/2)$, respectively, and respective time 2 prices $(4,4,16)$, $(4, 12, 4)$, $(4,2,2)$, $(2,2,6)$, $(2,4,2)$, $(2,1,1)$, $(1, 1/2, 1)$, $(1,1,1/2)$, $(1, 1/4, 1/4)$.*

1. *Show that this market is viable.*

2. *Find the martingale measure ν and the 2-forward measure λ, where S^0 is the chosen numeraire.*

3. *Find the (0,2)-forward price of S^1.*

4. *Find the 2-futures price of S^1 at time 0.*

CHAPTER 5

Incomplete Markets

In this chapter we pick up a thread left over from Chapter 3, namely: pricing claims in incomplete markets. An incomplete market has, by definition, non-attainable claims. For such claims (when there is no arbitrage), there are at least arbitrage enforced bounds on the price. Here we investigate this idea and characterize the arbitrage-allowed prices for non-attainable claims.

Theorem 3.7 says completeness is the same as uniqueness of the equivalent martingale measure, so an incomplete market admits multiple equivalent martingale measures. Let Σ denote the set of all possible equivalent martingale measures for a market model $(\Omega, \mu, \{\mathcal{F}_t\}, S)$ with numeraire S^0, and let ν_1 and ν_2 be two different elements of Σ. In a viable market, any attainable claim has a unique price at any time t. By Theorem 3.2, this is equal to the conditional expectation of the discounted payoff with respect to either ν_1 or ν_2.

However, for a non-attainable claim, these conditional expectations need not agree. The choice of a measure in Σ corresponds to the choice of a consistent (no-arbitrage) way to price all claims, even the non-attainable ones. We make this precise with the next definition. Recall that $\mathcal{X}^0$ is the set of claims attained by a self-financing strategy with initial value 0.

Definition 5.1 Consistent Pricing Scheme. A *consistent pricing scheme* Π is a linear functional

$$\Pi : \mathbb{R}^\Omega \to \mathbb{R}$$

such that

1. $\Pi(S_T^j) = S_0^j$ for all j,

2. whenever $X \geq 0$ and $\mu(X > 0) > 0$, then $\Pi(X) > 0$, and

3. $\Pi(X) = 0$ for all $X \in \mathcal{X}^0$.

In other words, a consistent pricing scheme is an assignment of a time-0 price to every claim in such a way that (1) the primary securities are priced correctly (and hence by linearity so are portfolios of the primary securities), (2) there are no arbitrage opportunities, and (3) the terminal payoff of a self-financing strategy with initial value zero is priced at zero.

Recall our notation $V_t(\phi) = \phi_t \cdot S_t$ for the time t value of a self-financing strategy ϕ. In the absence of arbitrage, this is the unique time-t price of the claim $X = \phi_T \cdot S_T$, as ϕ is a replicating strategy for X. We could denote this as $V_t(X)$, but we need a more general concept that works also when X is not attainable.

We introduce the notation

$$V_t(X, \nu) = E_\nu[(S_t^0/S_T^0)X|\mathcal{F}_t]$$

where X is any claim and ν is any equivalent martingale measure. When X is attainable, this is the unique price of X by Theorem 3.2. Otherwise, we now argue that $V_t(X, \nu)$ is a price consistent with no-arbitrage. More precisely, we have the following theorem.

Theorem 5.2

1. *For any equivalent martingale measure ν, the linear functional $\Pi(X) = V_0(X, \nu)$ is a consistent pricing scheme.*

2. *Conversely, for any consistent pricing scheme Π, there exists an equivalent martingale measure ν such that $\Pi(X) = V_0(X, \nu)$ for all X, namely, $\nu(A) = \Pi(S_T^0 \mathbf{1}_A)/\Pi(S_T^0)$ for $A \in \mathcal{F}$.*

Proof: For part 1, suppose we have an equivalent martingale measure ν, and define $\Pi(X) = V_0(X, \nu) = E_\nu[(S_0^0/S_T^0)X]$. This is evidently a linear functional on the space of claims; we need to verify the three conditions of the definition of consistent pricing scheme.

The first is immediate since S^j/S^0 is a ν-martingale for each j. The second follows since ν is equivalent to μ. The third condition follows from Theorem 3.2 and the definition of $\mathcal{X}^0$.

For part 2, given a consistent pricing scheme Π, let $\Pi_i = \Pi(\mathbf{1}_{\{\omega_i\}})$, where $\Omega = \{\omega_1, \ldots, \omega_M\}$ is the state space of size M.

Define $\lambda_i = \Pi_i / \sum \Pi_j$. Then $\lambda = (\lambda_1, \ldots, \lambda_M)$ is a positive vector in R^Ω orthogonal to $\mathcal{X}^0$ and also defines an equivalent probability measure since $\Pi_i > 0$ for all i.

Now we are in the same position as in the proof of Theorem 3.5. As before, we define the measure

$$\nu_i \equiv \nu(\omega_i) = \lambda_i S_T^0(\omega_i)/\lambda(S_T^0).$$

By the arguments in the proof of Theorem 3.5, ν is an equivalent martingale measure; it remains to verify that

$$\Pi(X) = V_0(X, \nu) \equiv E_\nu[X(S_0^0/S_T^0)]$$

for all claims X. To see this, first note that

$$\lambda(S_T^0) = \sum \lambda_i S_T^0(\omega_i) = \sum \Pi_i S_T^0(\omega_i)/\sum_j \Pi_j.$$

Now

$$\sum \Pi_i S_T^0(\omega_i) = \sum S_T^0(\omega_i)\Pi(\mathbf{1}_{\{\omega_i\}}) = \Pi(\sum S_T^0(\omega_i)\mathbf{1}_{\{\omega_i\}})$$

$$= \Pi(S_T^0) = S_0^0.$$

So $\lambda(S_T^0) = S_0^0 / \sum_j \Pi_j$ and

$$
\begin{aligned}
\Pi(X) &= \sum \Pi_i X(\omega_i) \\
&= \sum \Pi_i X(\omega_i) \left[\frac{S_0^0}{\lambda(S_T^0)} \frac{1}{\sum_j \Pi_j} \right] = \sum \lambda_i X(\omega_i) \frac{S_0^0}{\lambda(S_T^0)} \\
&= \sum \nu_i X(\omega_i) \frac{S_0^0}{S_T^0(\omega_i)} = E_\nu [X(S_0^0/S_T^0)].
\end{aligned}
$$

$\square$

This theorem means that the set Σ of all equivalent martingale measures is the same as the set of all ways to assign prices to all claims in a consistent, no-arbitrage way. This means that if we fix a claim X and let $\nu \in \Sigma$ vary, $V_0(X, \nu)$ ranges over the set

$$
\mathcal{P}(X) = \{ V_0(X, \nu) : \nu \in \Sigma \}
$$

of all possible "fair" prices of the claim in the sense of no-arbitrage.

Exercise 5.1 *Prove: the set Σ of equivalent martingale measures is **convex**: if ν_1 and ν_2 belong to Σ, then so does $\lambda \nu_1 + (1 - \lambda)\nu_2$ for any $\lambda \in [0, 1]$.*

Deduce that the set of possible prices $\mathcal{P}(X)$ for a fixed claim X is a convex subset of R. Since the convex subsets of R are the intervals, this means $\mathcal{P}(X)$ is an interval for every claim X.

For a claim X, the least upper bound and greatest lower bound of the interval $\mathcal{P}(X)$ are called the arbitrage-enforced bounds on the price of X. Any price in between these bounds is an allowed no-arbitrage price for X corresponding to some equivalent martingale measure, or equivalently to some consistent pricing scheme. Remember that when X is attainable, its price is unique and $\mathcal{P}(X)$ is just a single point. The remainder of this chapter is devoted to characterizing the interval $\mathcal{P}(X)$ when X is not attainable.

We say that a claim Y *dominates* a claim X if $Y \geq X$ with probability one. (In particular, every claim dominates itself.)

Theorem 5.3 *In a viable discrete market model, let X be a non-attainable claim, α be the supremum of all prices of attainable claims dominated by X, and β the infimum of all prices of attainable claims dominating X. Then $\mathcal{P}(X)$ is equal to the non-empty open interval (α, β).*

Proof: (This proof may be omitted on first reading.)
X is a real-valued function on the finite-dimensional Euclidean space $\mathbb{R}^\Omega$, so we can think of it equivalently as a vector $x \in \mathbb{R}^n$, where $n = |\Omega|$. Let C denote the closed positive cone in $\mathbb{R}^n$, and let $D = C \cup -C$ be the two-sided cone of all positive and all negative vectors. The translate $x + D$ of D is the collection of all claims either dominating or dominated by x.

Let V denote the subspace of $\mathbb{R}^n$ composed of all attainable claims. On V, we know there is a unique no-arbitrage linear pricing function $P : V \to \mathbb{R}$ given by the time-0 value of any replicating self-financing strategy. We discriminate reals p by the relation of the level set $\{P = p\} \equiv \{y \in V : P(y) = p\}$ with $x + D$. Define real subsets H, M, and L by

$$
\begin{aligned}
H &= \{p \in \mathbb{R} : \{P = p\} \cap (x + C) \neq \emptyset\} = P(V \cap (x + C)) \\
L &= \{p \in \mathbb{R} : \{P = p\} \cap (x - C) \neq \emptyset\} = P(V \cap (x - C)) \\
M &= \{p \in \mathbb{R} : \{P = p\} \cap (x + D) = \emptyset\} = \mathbb{R} \setminus (H \cup L)
\end{aligned}
$$

By construction, H is the set of prices of attainable claims dominating x, L is the set of prices of attainable claims dominated by x, and M is everything else.

Lemma 5.4   *Either $H \cap L = \emptyset$ or x is attainable.*

Proof: Suppose $p \in H \cap L$. Then there are non-negative vectors c_0 and c_1 such that both $x + c_0$ and $x - c_1$ are attainable, and $p = P(x + c_0) = P(x - c_1)$. Then $c_0 + c_1 = (x + c_0) - (x - c_1)$ is attainable and non-negative, with $P(c_0 + c_1) = 0$, so both c_0 and c_1 must be 0. $\square$

Lemma 5.5   *If $p \in H$, then $[p, \infty) \subset H$. If $p \in L$, then $(-\infty, p] \subset L$.*

Proof: Suppose $x + c \in (x + C) \cap V$ and $P(x + c) = p$. Then for any $\alpha \geq 0$, $x + c + \alpha S_T^0$ is also in $(x + C) \cap V$. Further, $P(x + c + \alpha S_T^0) = p + \alpha S_0^0$, which can attain any value $q \geq p$ for appropriate $\alpha \geq 0$. A similar argument applies for L. $\square$

Lemma 5.6   *The sets H and L are closed.*

Proof: Suppose we have a convergent sequence $p_n \in H$ with $p_n \to p$. There must be elements $c_n \in C$ such that, for all n, $x + c_n \in V$ and for any equivalent martingale measure ν,

$$
p_n = P(x + c_n) = V_0(x + c_n, \nu) = E_\nu[x \frac{S_0^0}{S_T^0}] + E_\nu[c_n \frac{S_0^0}{S_T^0}]. \tag{5.1}
$$

Now $c_n \in C$ must be a bounded sequence, or else the second term in the above sum would tend to infinity, contradicting the boundedness of the convergent sequence $\{p_n\}$. Therefore, $\{c_n\}$ has a convergent subsequence $\{c_{n_k}\}$ converging to a limit $c \in C$, and by continuity of P we have $P(x + c) = p$ so $p \in H$. This means H is closed. The proof for L is similar. $\square$

Lemma 5.7   *If ν is an equivalent martingale measure and $V_0(x, \nu) \in H$ or $V_0(x, \nu) \in L$ then x is attainable.*

Proof: If $V_0(x, \nu) \in H$, then there is a non-negative vector c such that $x + c \in V$ with $V_0(x, \nu) = V_0(x + c, \nu)$. Then $0 = V_0(0, \nu) = V_0(c, \nu)$, so $c = 0$, so $x = x + c \in V$. A similar argument applies for L. $\square$

From the definitions, it is clear that M is disjoint from both H and L. From the first lemma, assuming x is not attainable, then H, M, and L partition $\mathbb{R}$. From the second lemma, any value in H is at least as large as any value in M, and any value in L is no larger than any value in M, so all three sets are intervals. The third lemma implies that M is open. From the fourth lemma, we know that $\mathcal{P}(X) \subset M$, so M is non-empty. It remains to show that $M = \mathcal{P}(X)$.

Suppose that x is not attainable and $p \in M$. Since x is not in V, which is the domain of P, first define $Q : \operatorname{span}(x, V) \to \mathbb{R}$ by $Q(v + tx) = P(v) + tp$. Then Q is well-defined, linear on its domain, and agrees with P on V. If we can show that $\ker(Q) \cap D$ is exactly the origin, then we can determine a martingale measure ν exactly as in the proof of Theorem 3.5 so that $V_0(\cdot, \nu)$ extends Q, and it would follow that $p \in \mathcal{P}(X)$.

Lemma 5.8 *If a line in $\mathbb{R}^n$ passes through x, it is either entirely contained in $x + D$, or intersects $x + D$ only at x.*

Proof: A line through x consists exactly of points $x + tc$, for real t and a fixed vector c. The two cases are if $c \in D$ or not. $\square$

Fix a particular attainable claim v such that $P(v) = p$. Then an arbitrary claim in $\ker(Q)$ has a unique representation as $s(v - x) + z$ where $z \in \ker(P)$ and $s \in \mathbb{R}$. Assume $s \neq 0$, since $\ker(P)$ is already known to intersect D only at the origin. Adding x to this point in $\ker(Q)$ yields a point in the level set $\{Q = p\}$, re-written suggestively as

$$w = (1 - s)x + s(v + \frac{1}{s}z)$$

This is a certain point on the line that joins x to $v + \frac{1}{s}z \in \{P = p\}$. From the assumption that $p \in M$, this second point is not in $x + D$, so by the final lemma neither is w in $x + D$, so neither is $w - x = s(v - x) + z$ in D. $\square$

APPENDIX A

Probability Refreshner

We sketch some of the basic probability theory typically encountered in the context of mathematical finance, with an emphasis on the discrete case.

A.1   BEFORE PROBABILITY

We start with the *sample space* Ω, which is any old set. (In the body of this book, Ω is a finite set, but we don't restrict ourselves to the finite case for this chapter.) A point ω in Ω is a *sample point* or *outcome* or *state*.

An *event* is a subset of Ω, which is to say a collection of sample points. Not every subset is necessarily an event, but we insist that the collection $\mathcal{F}$ of all events satisfy the conditions of a σ-algebra, which we will state shortly.

There are some technical reasons for doing this, but also some very practical reasons. If one imagines the sample space Ω as a large set of all conceivable outcomes, then the σ-algebra $\mathcal{F}$ represents the limits of an observer's ability to distinguish between outcomes. For example, in a round of poker, a player initially sees only his own cards. This translates to a σ-algebra in which, for example, "I hold two pair" is an event, but "the opponent on my left holds a straight flush" is not. The situation is different for a spectator watching the same round of poker on television, who actually does know the initial hands of all players. This spectator's information translates to a different (larger) σ-algebra in which "player 1 holds two pair" and "player 2 holds a straight flush" are both events. In neither case is "the next card to be dealt is a heart" an event. (However, after some time, that next card will be dealt, so it is natural to consider a collection of σ-algebras indexed by time. This is a *filtration*, and we'll come back to that in a bit.)

We now fix a measurable space $(\Omega, \mathcal{F})$, where $\mathcal{F}$ is a σ-*algebra* of subsets of Ω. This means:

1. $\Omega \in \mathcal{F}$,

2. if $A \in \mathcal{F}$, then $A^c \in \mathcal{F}$, where A^c denotes the complement $\Omega \setminus A$ of A, and

3. if $\{A_i\}$ is a finite or countably infinite collection of sets in $\mathcal{F}$, then the union $\bigcup_i A_i \in \mathcal{F}$.

In brief, a sigma-algebra contains Ω and is closed under complementation and countable unions. This of course implies a sigma-algebra always contains the empty set and is closed under countable intersections as well.

Recall that a *partition* of Ω is a collection $\mathcal{A}$ of subsets of Ω such that any two elements of $\mathcal{A}$ are disjoint, and the union of all elements of $\mathcal{A}$ is Ω. A sigma-algebra $\mathcal{F}$ on Ω is *generated by a partition* if there is a partition $\mathcal{A}$ of Ω such that $\mathcal{A} \subset \mathcal{F}$ and $\mathcal{F}$ is the collection of all unions of elements of $\mathcal{A}$. The elements of $\mathcal{A}$ are called the *atoms* of $\mathcal{F}$.

Exercise A.1 *(a) If Ω is finite, show that any sigma-algebra $\mathcal{F}$ on Ω is generated by a partition.*

(b) More generally, show that any finite sigma-algebra on an arbitrary sample space is generated by a partition.

This $\mathcal{F}$ is the largest set of events we ever want to worry about, and it represents the ability of an omniscient observer to distinguish events. For finite sample spaces, $\mathcal{F}$ is usually taken to be the collection $\mathcal{P}(\Omega)$ of all subsets of Ω, though this is not required.

We will of course frequently consider sub-σ-algebras of $\mathcal{F}$.

Exercise A.2 *A sigma-algebra $\mathcal{F}$ distinguishes two outcomes $x \neq y \in \Omega$ if there exists an event $A \in \mathcal{F}$ containing exactly one element of $\{x, y\}$. Suppose Ω is a finite sample space and $\mathcal{F}$ is a sigma-algebra that distinguishes every pair of outcomes in Ω. Prove that $\mathcal{F}$ is the sigma-algebra $\mathcal{P}(\Omega)$ of all subsets of Ω.*

If $\Lambda \subset \mathbb{R}$ represents an index set of times, we say a *filtration* on Ω is a collection $\{\mathcal{F}_t : t \in \Lambda\}$ of sub-σ-algebras of $\mathcal{F}$ that is increasing with t. That is, if $s < t$ then $\mathcal{F}_s \subset \mathcal{F}_t$. A filtration represents an ability to distinguish sample points that does not degrade over time; therefore, it can be interpreted as describing the evolution of an observer's knowledge, as follows.

Consider the simplest case where the time index set Λ is finite, $\Lambda = \{0, \ldots, T\}$. An observer is watching an experiment unfold over time. At each time t a random outcome occurs, which is described by an event $A_t \subset \Omega$. The allowed events are determined by the filtration $\{\mathcal{F}_t\}$ from the requirements

1. for all $t \in \Lambda$, $A_t \in \mathcal{F}_t$, and

2. for all $s, t \in \Lambda$, if $s < t$, then $A_t \subset A_s$.

The decreasing sequence $\{A_t\}$ therefore represents increasingly more specific knowledge of which outcomes are consistent with all observations up to time t. In the case of finite Ω where $\mathcal{F}_T = \mathcal{P}(\Omega)$, A_T is then a single outcome in Ω, and there is no more uncertainty in the experiment. For $t < T$, the observer knows only that the outcome to be revealed by time T must lie in the event A_t but nothing more.

A measurable space with a filtration is called a *filtered measurable space*, denoted $(\Omega, \mathcal{F}, \{\mathcal{F}_t\})$, or simply $(\Omega, \{\mathcal{F}_t\})$ if $\mathcal{F}$ is understood or if $\mathcal{F} = \cup \mathcal{F}_t$.

A *random variable* X on Ω is a real-valued function $X : \Omega \to \mathbb{R}$. (Purists may complain that this should be called a statistic since we have not yet mentioned probability.) A random variable is a way of summarizing information about sample points into a single number. A random variable X may or may not summarize information in a way that respects the limitations determined by a σ-algebra $\mathcal{G}$. More formally, X is $\mathcal{G}$-*measurable* if for any real numbers a and b, the set

$$\{\omega \in \Omega \text{ such that } a \leq X(\omega) \leq b\}$$

is in $\mathcal{G}$. In case Ω is finite (and hence $\mathcal{G}$ is generated by atoms), this is equivalent to the simple requirement that X is constant on the atoms of $\mathcal{G}$. We always insist all random variables are $\mathcal{F}$-measurable.

Now consider an event A. This a collection of sample points, or in other words a subset of Ω, $A \subset \Omega$. We may define an associated random variable 1_A called the *indicator function for A* by the two exhaustive and exclusive cases $1_A(\omega) = 1$ if $\omega \in A$ and $1_A(\omega) = 0$ if $\omega \in \Omega \setminus A$. Pointwise multiplication of indicator functions corresponds to intersection of events, namely $1_{A \cap B}(\omega) = 1_A(\omega) 1_B(\omega)$. For a σ-algebra $\mathcal{F}$, the indicator 1_A is $\mathcal{F}$-measurable if and only if A is an element of $\mathcal{F}$.

The σ-algebra $\sigma(X)$ *generated by a random variable X* is the smallest σ-algebra for which X is $\sigma(X)$-measurable. For the indicator function, note that $\sigma(1_A) = \{\emptyset, A, \Omega - A, \Omega\}$.

Exercise A.3 *Suppose Ω is finite and X is a random variable on $(\Omega, \mathcal{F})$. Show that another random variable Y is $\sigma(X)$-measurable if and only if there is a function $g : \mathbb{R} \to \mathbb{R}$ such that $Y = g(X)$.*

Given a filtration $\{\mathcal{F}_t\}$ and a stochastic process $\{X_t\}$, we say $\{X_t\}$ *is $\{\mathcal{F}_t\}$-adapted* if each random variable X_s is $\mathcal{F}_t$-measurable whenever $s \leq t$. Intuitively, an adapted process is one whose value at any time t is actually observable at that time, given the information restrictions of the filtration.

Given a stochastic process $\{X_t\}$, there is a smallest filtration $\{\mathcal{G}_t\}$ to which it is adapted. Note that $\mathcal{G}_t$ is not generally just $\sigma(X_t)$, since X_s also needs to be $\mathcal{G}_t$-measurable.

If $S_t^0, \ldots, S_t^k$ are all random processes adapted to $\{\mathcal{F}_t\}$, then we say the vector process $S_t = (S_t^0, \ldots, S_t^k)$ is adapted to $\{\mathcal{F}_t\}$. We usually use the notation $S = \{S_t : t \in \Lambda\}$ for this vector process. Interpreting the components of S as asset prices, we use the notation $(\Omega, \mu, \{\mathcal{F}_t\}, S)$ to designate a market model specifying a space of outcomes Ω, a reference probability measure μ (see the next section), an information structure described by the filtration $\{\mathcal{F}_t : t \in \Lambda\}$, and the asset price processes S.

The subject of this book is *discrete* market models $(\Omega, \mu, \{\mathcal{F}_t\}, S)$, meaning that the index set Λ of times is finite, $\Lambda = \{0, \ldots, T\}$, and the sigma-algebra $\mathcal{F} = \mathcal{F}_T$ is finite. By identifying atoms of $\mathcal{F}$ as single points, we can obtain a finite measurable space that is equivalent to the original. Therefore, there is no loss of generality by simply taking our discrete models to be finite to begin with.

A.2   PROBABILITY INTRODUCED; INDEPENDENCE

Given a measurable space $(\Omega, \mathcal{F})$, a *probability law* (or *probability measure*) is a non-negative countably additive function P on $\mathcal{F}$, with $P(\Omega) = 1$. More explicitly:

1. $0 \leq P(A) \leq 1$ for all $A \in \mathcal{F}$,

2. $P(\Omega) = 1$, and

3. if $\{A_i\}$ is a disjoint sequence of sets in $\mathcal{F}$ and the union $\cup A_i$ belongs to $\mathcal{F}$, then

$$P(\cup_i A_i) = \sum_i P(A_i).$$

The triple $(\Omega, P, \mathcal{F})$ is called a *probability space*. Probabilists often have only one probability law in mind and start with a probability space as the point of origin. We definitely plan to change the probability law, so for us, the point of origin is a measurable space $(\Omega, \mathcal{F})$ or a filtered measurable space $(\Omega, \{\mathcal{F}_t\}, \mathcal{F})$. Choosing a measure P then determines a probability space $(\Omega, P, \mathcal{F})$ or a *filtered probability space* $(\Omega, P, \{\mathcal{F}_t\}, \mathcal{F})$.

Note: Beginners sometimes get confused on the following point. P assigns a value between 0 and 1 to an event. P is not a random variable. However, if we fix a P and then think about another Q, which Q is also not a random variable, then we can describe the relation taking P to Q by a random variable. This is summarized by the Radon-Nikodym Theorem, below.

Given a random variable $X : \Omega \to \mathbb{R}$ we define its *distribution*, or *cumulative distribution function*, $F_X : \mathbb{R} \to [0, 1]$ by

$$F_X(x) = P(\{\omega \in \Omega \text{ such that } X(\omega) \le x\})$$

With a probability law P in mind, the expectation $E(X) \equiv E_P(X)$ of a random variable X is defined to be

$$E(X) = \int X dP = \int_\Omega X(\omega) P(d\omega) = \int_R x \, dF_X(x)$$

and in case Ω is finite and $\mathcal{F}$ is the power set of Ω (that is, generated by the singletons), this becomes

$$E(X) = \sum_{\omega \in \Omega} X(\omega) P(\{\omega\}).$$

We also make use of the convenient notation $P(X) = E_P(X)$ even though this is a slight abuse, since P started out as a set function. The two notations are connected by $P(\mathbb{1}_A) = E_P(\mathbb{1}_A) = P(A)$, where $\mathbb{1}_A$ is the indicator function of the set A. Context determines whether the measure P is being used as a set function or a function of random variables.

Exercise A.4 *Thinking of the probability measure P as a function of random variables as above, verify that P is a linear functional. That is, $P : \mathbb{R}^\Omega \to \mathbb{R}$ is linear in the sense that, for all $c \in \mathbb{R}$ and $X, Y \in \mathbb{R}^\Omega$,*

1. *$P(cX) = cP(X)$ and*

2. *$P(X + Y) = P(X) + P(Y)$.*

We say that X is *integrable* if $E(|X|) < \infty$, and notice that this is automatic when Ω is finite. If A and B are two events and $P(A) > 0$, then we can define the *conditional probability*

$$P(B|A) = \frac{P(A \cap B)}{P(A)}.$$

Intuitively, this is the probability of event B occurring given the knowledge that event A has occurred.

A and B are *independent* if $P(A \cap B) = P(A)P(B)$, which is equivalent to requiring $P(B|A) = P(B)$, or $P(A|B) = P(A)$. A finite collection $\mathcal{A} = \{A_1, \ldots, A_n\}$ is said to be independent if, for any subcollection of distinct sets in $\mathcal{A}$, the probability of their intersection is equal to the product of their probabilities. Then we say that an infinite collection $\mathcal{A}$ of sets is independent if every finite subcollection is.

Now given classes of sets $\mathcal{A}_1, \ldots, \mathcal{A}_n$, we say they are independent if for each choice of $A_i \in \mathcal{A}_i, i = 1, \ldots, n$, the events $A_1, \ldots, A_n$ are independent.

At last we are able to say that the random variables $X_1, \ldots, X_n$ are independent if the sigma-algebras $\sigma(X_1), \ldots, \sigma(X_n)$ are independent.

A.3   CONDITIONAL EXPECTATION – THE FINITE CASE

We start with an integrable random variable X. First define expectation of X conditional on an event A as the real number

$$E_P(X|A) = E_P \left(\frac{1_A X}{P(A)} \right)$$

The 1_A factor says to ignore any sample point not in A, and the $1/P(A)$ re-inflates the total mass to 1. This ties back to conditional probability by

$$E_P(1_B|A) = P(B|A)$$

using once again the close relation between an event B and its indicator function 1_B. If A_i are disjoint, then immediately

$$\sum_i E_P(X|A_i)P(A_i) = E_P(X|\bigcup_i A_i)P(\bigcup_i A_i)$$

This is especially interesting when the A_i partition the sample space Ω, so that the right-hand side is the unconditional expectation $E_P(X)$ Recalling that $P(A_i) = E(1_{A_i})$, it's tempting to look at the left-hand side as the expectation of the certain random variable

$$\sum_i E_P(X|A_i)1_{A_i}$$

This random variable is a handy way to simultaneously keep track of all the numbers $E_P(X|A_i)$. It is not X, but, in a sense to be made precise later, it is the closest approximation to X in terms of functions generated by the indicators 1_{A_i}. Note that the functions generated by the 1_{A_i} are exactly the functions measurable with respect to the σ-algebra $\sigma(\{A_i\})$ generated by the A_i.

We now define the conditional expectation $E(X|\mathcal{G})$ when $\mathcal{G}$ is a finite sigma-algebra, for example in the case of finite sample space Ω. By Exercise A.1, there is a finite partition $\{A_i\}$ of Ω such that $\mathcal{G} = \sigma(\{A_i\})$, and we define the conditional expectation

$$E_P(X|\mathcal{G}) \equiv E_P(X|\sigma(\{A_i\})) = \sum_i E_P(X|A_i)1_{A_i}$$

and clearly this is automatically $\mathcal{G}$-measurable.

As a special case, suppose $B \subset A_j$ for some j. Then

$$E_P(1_B|\sigma(\{A_i\})) = \sum_i E_P(1_B|A_i)1_{A_i} = \frac{P(B)}{P(A_i)}1_{A_i}.$$

Exercise A.5 *Show that $E_P(XY|\sigma(\{A_i\})) = X E_P(Y|\sigma(\{A_i\}))$ when X is already measurable with respect to $\sigma(\{A_i\})$.*

Exercise A.6 *Tower property: If $\mathcal{G} \subset \mathcal{F}$ are two finite sigma-algebras, then*

$$E(E(X|\mathcal{F})|\mathcal{G}) = E(X|\mathcal{G}). \tag{A.1}$$

In the special case where $\mathcal{G}$ is the trivial sigma-algebra $\{\emptyset, \Omega\}$, then $E(X|\mathcal{G}) = E(X)$ and

$$E(E(X|\mathcal{F})) = E(X). \tag{A.2}$$

Conditional expectations may be thought of as orthogonal projections with respect to the L^2 inner product on random variables defined by

$$\langle X, Y \rangle = \int XY \, dP = E_P(XY).$$

Exercise A.7 *Prove the statement above by showing that $\langle X - E(X|\mathcal{F}), Y \rangle = 0$ for every $\mathcal{F}$-measurable Y, and therefore $E(X|\mathcal{F})$ is the projection of X onto the linear subspace of $\mathcal{F}$-measurable random variables.*

A.4   CHANGE OF MEASURE; THE RADON-NIKODYM DERIVATIVE

If P and Q are probability measures on $(\Omega, \mathcal{F})$, then Q *is absolutely continuous with respect to* P, written $Q << P$, if $Q(A) = 0 \Rightarrow P(A) = 0$ for any event A. Intuitively, this means that Q does not have any mass in any place that P doesn't also have mass.

Theorem A.1 *If $Q << P$, then for any $\delta > 0$, there exists $\epsilon > 0$ such that if $P(A) < \epsilon$ then $Q(A) < \delta$.* The proof is easy and in Billingsley [Bil95] or the green Rudin [Rud74]. The similarity of this

statement to the ϵ-δ definition of continuity of functions of a real variable suggests how the property may have got its name.

Theorem A.2   Radon-Nikodym. *Suppose Q and P are probability measures on $(\Omega, \mathcal{F})$, and that $Q << P$. Then there is an $\mathcal{F}$-measurable random variable $g : \Omega \to \mathbb{R}$ such that for any event A, we have $Q(A) = E_P(g \cdot 1_A)$.*

This is an important theorem in measure theory and is easy to prove when Ω is finite, as follows. Let $\{A_i\}$ be a finite partition generating $\mathcal{F}$. Define the random variable

$$g = \sum_i \frac{Q(A_i)}{P(A_i)} 1_{A_i}$$

If $A \in \mathcal{F}$ is any event, then $A = \cup_{i \in J} A_i$ for some finite index set J, and so

$$E_P(g \cdot 1_A) = E_P(\sum_{i \in J} \frac{Q(A_i)}{P(A_i)} 1_{A_i}) = \sum_{i \in J} \frac{Q(A_i)}{P(A_i)} E_P(1_{A_i}) = Q(A).$$

The customary notation for g is $\frac{dQ}{dP}$, and is called the *Radon-Nikodym derivative of Q with respect to P*. One can think of this random variable $\frac{dQ}{dP}$ as a way to express expectations with respect to Q in terms of expectations with respect to P via the formula

$$E_Q(X) = E_P(X\frac{dQ}{dP}) \tag{A.3}$$

Exercise A.8 *Verify Equation* (A.3).

The Radon-Nikodym derivative also provides us with a *change of measure formula for conditional expectation*. For any integrable random variable X, and a sub-sigma-algebra $\mathcal{G}$ of $\mathcal{F}$, if $Q << P$,

$$E_Q(X|\mathcal{G}) = \frac{E_P(X\frac{dQ}{dP}|\mathcal{G})}{E_P(\frac{dQ}{dP}|\mathcal{G})}. \tag{A.4}$$

Exercise A.9 *For $\mathcal{F}$ finite, directly verify equation* (A.4).

An important notion in the book is the concept of equivalent measures. If P and Q are two probability measures on the measurable space $(\Omega, \mathcal{F})$, then we say *P is equivalent to Q, $P \sim Q$*, if $P << Q$ and $Q << P$. This means that the two measures agree on which sets have measure zero and which sets have positive measure. This defines an equivalence relation on measures, and an *equivalence class* of measures is the collection of all measures equivalent to some particular measure; as usual, the set of all equivalence classes forms a partition of the set of all probability measures on $(\Omega, \mathcal{F})$.

Exercise A.10 *Verify that equivalence of measures is an equivalence relation. That is, for any probability measures P, Q, R on $(\Omega, \mathcal{F})$,*

1. $P \sim P$,

2. if $P \sim Q$ then $Q \sim P$, and

3. if $P \sim Q$ and $Q \sim R$, then $P \sim R$.

In the study of market models $(\Omega, \mu, \{\mathcal{F}_t\}, S)$, the role of the reference measure is merely to determine an equivalence class of measures. That is, the measure μ determines which subsets of Ω have measure zero, but no other information about μ is needed. This becomes even simpler for our finite models, because we take the convention that every element of Ω has positive μ-measure. Therefore, the measures equivalent to μ are just those that have the same property.

A.5   MARTINGALES

Roughly speaking, a martingale is a random process whose conditional expectation at any future time is its current value – it is constant on average. This concept is central to the field of asset pricing and probability more generally. For our purposes, we only need the basic definition here. The financial significance of martingales is one of the themes developed in the main body of this text.

Definition A.3   Given a filtered measurable space $(\Omega, \{\mathcal{F}_t\}, \mathcal{F})$ and an adapted stochastic process Y_t, we say that the measure ν on $(\Omega, \mathcal{F})$ is a *martingale measure for Y_t*, or that Y_t *is a ν-martingale*, if Y_t is integrable with respect to ν for each t, and

$$Y_s = E_\nu(Y_t | \mathcal{F}_s) \text{ for all } s, t \text{ with } 0 \le s \le t \le T. \tag{A.5}$$

Exercise A.11 *Show that the each of the following is equivalent to the condition* (A.5):

1. $Y_t = E_\nu(Y_{t+1} | \mathcal{F}_t)$ *for all* $t = 0, \ldots, T - 1$.

2. $Y_t = E_\nu(Y_T | \mathcal{F}_t)$ *for all* $t = 0, \ldots, T$.

Exercise A.12 *If X is an integrable random variable on $(\Omega, \nu, \{\mathcal{F}_t\}, \mathcal{F})$, then the process X_t defined by*

$$X_t = E_\nu(X | \mathcal{F}_t)$$

is a ν-martingale.

APPENDIX B

Orthogonal Vectors in the Positive Cone

The purpose of this appendix is to give an elementary and self-contained proof of proposition B.1, used in the proof of the second fundamental theorem of asset pricing, Theorem 3.5.

In $\mathbb{R}^n$, let P denote the open positive cone, that is, the set of vectors all of whose coordinates are positive. Let P^* denote the closure of P minus the origin, that is, the set of nonzero vectors all of whose coordinates are non-negative.

Let L be any linear subspace of $\mathbb{R}^n$.

Proposition B.1

(a) If $L \cap P^* = \emptyset$, then there exists $v \in P$ such that v is orthogonal to L.

(b) If $L \cap P = \emptyset$, then there exists $v \in P^*$ such that v is orthogonal to L.

The proof requires two lemmas.

Lemma B.2   *Let C be a closed convex set in $\mathbb{R}^k$, and suppose $y \notin C$. Then there exists $v \in \mathbb{R}^k$ such that $x \cdot v > y \cdot v$ for all $x \in C$.*

Proof: Choose $z \in C$ having minimum distance to y. (This z is unique, but we don't need that fact.) Let $v = z - y$. For any $x \in C$, convexity implies that $z + \lambda(x - z) \in C$ for all $\lambda \in (0, 1]$. Therefore,

$$||z + \lambda(x - z) - y||^2 \geq ||z - y||^2.$$

Expanding and simplifying gives

$$2\lambda(x - z) \cdot z - 2\lambda(x - z) \cdot y + \lambda^2 ||x - z||^2 > 0.$$

Dividing by 2λ and letting $\lambda \to 0$ gives $(x - z) \cdot z - (x - z) \cdot y \geq 0$ or, equivalently, $x \cdot v \geq z \cdot v$ for all $x \in C$.

Also, we have $0 < ||v||^2 = (z - y) \cdot v = z \cdot v - y \cdot v$, so $z \cdot v > y \cdot v$. Combining these two inequalities gives our result. □

Lemma B.3   *Let L be a linear subspace of $\mathbb{R}^n$, $C \subset L$ be a convex set, and $0 \notin C$. Let $\bar{C}$ denote the closure of C.*

Then there exists a nonzero vector $v \in L$ such that for all $x \in \bar{C}, x \cdot v \geq 0$.

Proof: For $k = \dim(L)$, L is isometrically isomorphic to $\mathbb{R}^k$, so it suffices to consider the case $L = \mathbb{R}^k$. The set $\bar{C}$ is convex. Since $0 \notin C$, there is a sequence $\{y_n\}$ of points outside $\bar{C}$ such that $y_n \to 0$.

Apply Lemma B.2 to each y_n to get $v_n \in L$ such that $x \cdot v_n > y_n \cdot v_n$ for all for all $x \in \bar{C}$. Since $v_n \neq 0$, we may assume by scaling that it is a unit vector. Choosing a convergent subsequence $v_{n_i} \to v \neq 0$ and letting $i \to \infty$ gives the desired conclusion. $\square$

Proof of Proposition B.1:

(a) Let π be the orthogonal projection of $\mathbb{R}^n$ onto L. We wish to show there is $v \in P$ such that $\pi(v) = 0$. Assume for contradiction that there is no such v.

Then $Q \equiv \pi(P)$ does not contain the origin. Further, since P is convex, so is Q. By Lemma B.3, there is a nonzero $h \in L$ such that for all $x \in Q$, $x \cdot h \geq 0$.

Since h cannot be in P^*, it must have a negative component. Equivalently, there is a standard basis vector e of $\mathbb{R}^n$ such that $e \cdot h < 0$. By continuity, there is $f \in P$ near e such that $f \cdot h < 0$. So $\pi(f) \cdot h = f \cdot h < 0$. Since $\pi(f) \in Q$, we also have $\pi(f) \cdot h > 0$ by above, a contradiction.

(b) If $L \cap P = \emptyset$, the L is a limit of subspaces L_j of the same dimension such that $L_j \cap P^* = \emptyset$ for each j. Applying part (a) of this proposition to L_j gives us a vector $v_j \in P$ orthogonal to L_j. We may as well assume v_j is a unit vector. Then some subsequence converges to a nonzero vector $v \in P^*$ orthogonal to L. $\square$

Bibliography

[Bil95] P. Billingsley. *Probability and Measure*. Wiley, third edition, 1995. xi, 44

[BR96] M. Baxter and A. Rennie. *Financial Calculus*. Cambridge University Press, Cambridge, UK, 1996. x

[BS73] F. Black and M. Scholes. The pricing of options and corporate liabilities. *Journal of Political Economy*, 81:637–654, 1973. DOI: 10.1086/260062 x

[Chu74] K. L. Chung. *A course in probability*. Academic Press, second edition, 1974. xi

[CRR79] J. Cox, S. Ross, and M. Rubinstein. Option pricing: a simplified approach. *Journal of Financial Economics*, 7:229–263, 1979. DOI: 10.1016/0304-405X(79)90015-1 ix

[DS94] F. Delbaen and W. Schachermayer. A general version of the fundamental theorem of asset pricing. *Mathematische Annalen*, 300:463–520, 1994. DOI: 10.1007/BF01450498 x

[DS98] F. Delbaen and W. Schachermayer. The fundamental theorem of asset pricing for unbounded stochastic processes. *Mathematische Annalen*, 312:215–250, 1998. DOI: 10.1007/s002080050220 x

[DS06] F. Delbaen and W. Schachermayer. *The Mathematics of Arbitrage*. Springer Finance. Springer-Verlag, Berlin, 2006. x

[Duf96] D. Duffie. *Dynamic Asset Pricing Theory*. Princeton University Press, Princeton, NJ, second edition, 1996. x

[EK99] R. J. Elliot and P. Ekkehard Kopp. *Mathematics of Financial Markets*. Springer-Verlag, New York, 1999. x, 27

[HK79] M. Harrison and D. Kreps. Martingales and arbitrage in multi-period securities markets. *Journal of Economic Theory*, 20:381–408, 1979. DOI: 10.1016/0022-0531(79)90043-7 x

[HP81] M. Harrison and S. Pliska. Martingales and stochastic integrals in the theory of continous trading. *Stochastic Processes and their Applications*, 11:215–260, 1981. DOI: 10.1016/0304-4149(81)90026-0 x

[Hul08] J. Hull. *Options, Futures, and Other Derivatives*. Prentice Hall, New Jersey, seventh edition, 2008. x

[JP00] J. Jacod and P. Protter. *Probability Essentials*. Springer-Verlag, New York, 2000. xi

[Kre81] D. Kreps. Arbitrage and equilibrium in economies with infinitely many commodities. *Journal of Mathematical Economics*, 8(1):15–35, 1981. DOI: 10.1016/0304-4068(81)90010-0 x

[Mer73] R. C. Merton. Theory of rational option pricing. *Bell Journal of Economics and Management Science*, 4:141–183, 1973. DOI: 10.2307/3003143 x

[Pli97] S. Pliska. *Introduction to Mathematical Finance: Discrete Time Models*. Blackwell, Oxford, 1997. x

[Rud74] W. Rudin. *Real and Complex Analysis*. McGraw-Hill, New York, second edition, 1974. 44

[Shr04] S. Shreve. *Stochastic Calculus for Finance I: The Binomial Asset Pricing Model*. Springer Finance. Springer-Verlag, New York, 2004. x

Authors' Biographies

GREG ANDERSON

Greg Anderson is a director in the rates quant group at Bank of America Merrill Lynch. Previously, he was in the fixed income research group at MSCIBarra. He holds a PhD in mathematics from the University of California at Berkeley.

ALEC N. KERCHEVAL

Alec N. Kercheval is professor of mathematics at Florida State University, where he has been director of the financial mathematics graduate program. He earned a PhD in mathematics from the University of California, Berkeley, and has taught at Boston University, Indiana University, Bloomington and the University of Texas at Austin. Prior to arriving at Florida State University, he worked in the fixed income research group at MSCIBarra, a financial consulting firm in Berkeley, California.